The Roman Army of the Middle Empire, AD 180–284

Gabriele Esposito is a military historian who works as a freelance author and researcher for some of the most important publishing houses in the military history sector. In particular, he is an expert specializing in uniformology: his interests and expertise range from the ancient civilizations to modern post-colonial conflicts. During recent years, he has conducted and published several researches on the military history of the Latin American countries, with special attention on the War of the Triple Alliance and the War of the Pacific. He is among the leading experts on the military history of the Italian Wars of Unification and the Spanish Carlist Wars. His books and essays are published on a regular basis by Osprey Publishing, Winged Hussar Publishing and Libreria Editrice Goriziana. He is also the author of numerous military history articles appearing in specialized magazines such as *Ancient Warfare Magazine*, *Medieval Warfare Magazine*, *The Armourer*, *History of War*, *Guerres et Histoire*, *Focus Storia* and *Focus Storia Wars*.

The Roman Army of the Middle Empire, AD 180–284

Weapons, Organization and Equipment

Gabriele Esposito

Pen & Sword
MILITARY

First published in Great Britain in 2024 by
Pen & Sword Military
An imprint of
Pen & Sword Books Limited
Yorkshire – Philadelphia

ISBN 978 1 39903 181 3

A CIP catalogue record for this book is available from the British Library

Typeset by Mac Style
Printed and bound in India by Parksons Graphics Pvt. Ltd.

Pen & Sword Books Limited incorporates the imprints of After the Battle, Atlas, Archaeology, Aviation, Discovery, Family History, Fiction, History, Maritime, Military, Military Classics, Politics, Select, Transport, True Crime, Air World, Frontline Publishing, Leo Cooper, Remember When, Seaforth Publishing, The Praetorian Press, Wharncliffe Local History, Wharncliffe Transport, Wharncliffe True Crime and White Owl.

For a complete list of Pen & Sword titles please contact:

PEN & SWORD BOOKS LIMITED
47 Church Street, Barnsley, South Yorkshire, S70 2AS, England
E-mail: enquiries@pen-and-sword.co.uk
Website: www.pen-and-sword.co.uk
or
PEN AND SWORD BOOKS
1950 Lawrence Rd, Havertown, PA 19083, USA
E-mail: uspen-and-sword@casematepublishers.com
Website: www.penandswordbooks.com

Contents

Acknowledgements vi
Introduction vii

Chapter 1 The Crisis of the Third Century 1

Chapter 2 The Roman Army of the Third Century 14

Chapter 3 The Wars of the Middle Empire 53

Chapter 4 The Military Reforms of Constantine 88

Chapter 5 Weapons and Tactics 121

Bibliography 187
The Re-enactors who Contributed to this Book 189
Index 195

Acknowledgements

This book is dedicated to my magnificent parents, Maria Rosaria and Benedetto, for the immense love and fundamental support that they always give me. Thanks to their precious advice deriving from long experience, the present book is much better than planned. A very special thanks goes to Philip Sidnell, the commissioning editor of my books for Pen & Sword: his love for history and his passion for publishing are the key factors behind the success of our publications. Many thanks also to the production manager of this title, Matt Jones, for his excellent work and great enthusiasm. A special thanks is due to Tony Walton for editing my original manuscript in a spectacular way. A very special mention goes to the brilliant reenactment groups that collaborated with their photos to the creation of this book: without the incredible work of research of their members, in fact, the final result of this publication would have not been the same. As a result, I want to express my deep gratitude to the following living history associations/single reenactors: 'Ala I Batavorum' from the Netherlands (especially in the person of Jurjen Draaisma), Conrad Lunar from Germany, 'Legio VI Victrix Cohors II Cimbria' from Scandinavia, 'Legio XI CPF' from Switzerland (especially in the person of Yves Ruttimann), 'Numerus Brittonum' from Germany, Paul Elliott from the United Kingdom, 'Protectores Domini Nostri' from Italy and 'Varianus' from Germany (especially in the person of Robin Rechkemmer).

Introduction

During the years of the Middle Empire (AD 180–284), the Roman Army had to face the terrible Crisis of the Third Century. The latter was the result of many different factors, including gigantic socio-economic problems and the arrival of new Germanic populations on the borders of the Roman Empire. The Roman state, after the flourishing years of the Early Empire, appeared at risk of disappearing two centuries before the actual fall of the Western Roman Empire in AD 476. Several important provinces became temporarily independent from the central government of Rome, while large areas of the Empire where raided by foreign invaders. The Roman Army had to fight with all its resources in order to reconquer a large portion of the Empire and preserve its unity. The only way to survive for the Roman armies was to modernize and reform their main structures in order to face the new threats posed by a multitude of warlike enemies, from the Persians in the Middle East to the Germanic peoples in Central Europe. The previous military system based on the power of the legions' heavy infantry was completely revised, with the introduction of new organizational patterns. Cavalry became much more important than before, together with light troops specialized in skirmishing. The personal equipment of the Roman soldiers also changed very significantly, with the introduction of new weapons that permitted the development of innovative tactics. In this book we will follow the evolution of the Roman Army from the death of Marcus Aurelius to the rise of Diocletian, reconstructing the major military campaigns fought from AD 180–284 and explaining how Rome's military forces entered the Late Empire.

Chapter 1

The Crisis of the Third Century

After several decades of internal prosperity and military expansionism on its borders, the Roman Empire started to experience a general crisis following the death of Marcus Aurelius in AD 180 that affected all its structures, including its military. This turbulent period of Roman history, commonly known as the Crisis of the Third Century, lasted from 235 until the ascension of Diocletian in 284. Conventionally, the assassination of Emperor Severus Alexander is considered the starting point of the crisis, but it should be pointed out that the internal and external troubles of the Roman Empire had started long before 235. The first signs of military difficulties, for example, had become apparent during the reign of Marcus Aurelius: the terrible Marcomannic Wars (166–180) had already shown that the Germanic tribes of Northern and Central Europe were becoming a serious menace for the stability of the Roman frontiers. Since the reign of Augustus, the Romans had abandoned their ambitions to occupy Germany, but had been able to contain the Germanic peoples and to effectively deal with their incursions. The *limes* – the border between the Roman Empire and the barbarians – had remained quiet and stable for many decades, the military supremacy of the Romans having never been contested and the relations with the Germani having generally remained positive. During the second half of the second century AD, however, this situation changed in a dramatic way: coming under pressure from warlike peoples migrating from the steppes of Central Asia and Eastern Europe, the Germani started to move towards the borders of the Roman Empire. For the first time in their recent history, the Romans had to face massive migrations of entire peoples, not just the minor raids of a single tribe. The defensive system based on garrisons of legions and locally recruited auxiliaries had no hope of countering the massive penetrations of the Germanic peoples now on the move. The border was too vast to be defended effectively, with simultaneous attacks from several different tribes becoming common. The Germanic peoples were a confederation of tribes, which could assemble large numbers of warriors for attacks against particular sectors of the Roman *limes*.

During the Marcomannic Wars, the Romans experienced various military difficulties, and although they were able to defeat the Germani, they could only do so with incredible difficulty and after suffering severe losses. To this we should add the

Antonine Plague that killed thousands of individuals across the Empire, as well as the constant state of war that existed on Rome's eastern borders, where the Parthians were always active militarily. Thanks to the leadership of Marcus Aurelius, the Roman Empire was able to survive the difficult period of the Marcomannic Wars, but many questions remained dangerously unresolved. The immediate successors of Marcus Aurelius, starting from his son Commodus, did very little to improve the general situation of the Empire's defences. With the ascendancy of the new Severan Dynasty, the institutional nature of the Roman Empire was changed: beginning with Septimius Severus, the Roman emperors began to be military despots, only able to retain power thanks to the decisive support of the army and the equestrian social class. The traditional respect that the emperors had always shown towards the Senate started to disappear, the ruler enjoying absolute power and being able to impose his will by the use of violence. The senatorial class gradually lost importance, especially in the military: loyalty towards the supreme leader became increasingly more important than ability or virtue. Since the soldiers were the real holders of power, emperors started to treat the armies as their own private property. Large donatives of money were distributed to the legionaries, the pay was gradually augmented and the number of troops under direct control of the emperor was increased. Septimius Severus was the first monarch to garrison a legion near Rome. It should also be pointed out that, after winning the civil war that gave him the Empire, Septimius Severus replaced all the Praetorians with new men from his loyal legions stationed in Pannonia (men who had shown their loyalty to him during the recent internecine fighting).

Another great political change, happening in the eastern part of the Empire, was to have very important consequences for the military situation of the Romans. In 228, the Parthians, who had controlled present-day Iran since the fall of the Seleucid Empire, were decisively defeated by a new Persian dynasty that assumed control of their territories. The new rulers of Iran were the Sassanids, who soon massively transformed the former Parthian Empire. The Parthians, coming from the steppes of Central Asia, had created a feudal state that was a confederation of several regions that were largely autonomy. However, the Sassanids soon reorganized the empire as a centralized nation with a strong ruler and single religion. In doing so, they presented themselves as the heirs of the ancient Achaemenids, who had created a solid and immense multinational empire. As a result of all these changes, the eastern part of the Roman Empire started to be menaced by a new threat that was much more dangerous than the Parthians. The Sassanids had a political organization that was in many respects very similar to that of Rome, and could confront the Roman Empire on almost equal terms. While the Parthians had generally limited themselves to raids and guerrilla campaigns, the Sassanids were capable of launching full-scale invasions

Legionary from the Early Empire, equipped with *lorica segmentata* segmented cuirass. (*Photograph by Simone Ochsner (https://www.simoneochsner.com), copyright Legio XI CPF*)

Legionary from the Early Empire, armed with *pilum* throwing javelin. (*Photograph by Simone Ochsner (https://www.simoneochsner.com), copyright Legio XI CPF*)

Legionary from the Early Empire, attacking with *gladius* short sword. (*Photograph by Simone Ochsner (https://www.simoneochsner.com), copyright Legio XI CPF*)

of the Rome's rich eastern provinces. Since the ascendancy of the new dynasty, a state of war on the eastern *limes* was almost continuous. In 260, Emperor Valerian was defeated in battle and captured by the Sassanids, beginning one of the worst moments in the long history of Rome. With no more central authority to speak of, the Empire risked disappearing forever as a result of internal secessions and external invasions. Many thousands of Germanic warriors now crossed the Roman borders and laid waste to large areas of the northern provinces, while the Sassanids found no opposition in their occupation of major parts of the eastern territories. At this point, to enable it to effectively face the foreign invasions, two large parts of the Empire abandoned Rome's central government and started to act as independent political entities. The territories of modern France, Britain and Spain proclaimed their independence as the Gallic Empire, while most of the eastern provinces were absorbed into the Kingdom of Palmyra.

Until the rise of the Sassanids, the desert city of Palmyra had been a small client state of Rome, having great commercial importance but being very limited politically. The defeat of Valerian, however, quickly changed matters, with Odenatus, the astute King of Palmyra, taking the opportunity of this political vacuum to enlarge his territories. He presented himself as the legitimate defender of Roman interests in the East because of his alliance with Rome, thanks to which he was able to unite most of the eastern provinces under his direct control and launch a strong counter-offensive against the Sassanids. Judging the secessions of the third century from a historical point of view, it can be said that it was only thanks to them that the Roman Empire was able to survive. The division of the imperial territories enabled the local leaders and military forces to put up a better defence against the foreign invasions. This was the case with Gaul, where the usurper Postumus was able to defeat the Germani, while more or less the same happened in Syria and Egypt, where Odenatus contained the Sassanid advances. The Gallic Empire remained an independent state during the period 260–274, while the Kingdom of Palmyra acted as a free country from 261 until 272. Both the secessions, however, were decisively crushed by Emperor Aurelianus, who was also able to successfully fight Rome's external enemies.

The entire period of the Crisis of the Third Century is also known as the Military Anarchy due to the total chaos in the succession of emperors caused by the increasing power of the military. From the reign of Septimius Severus, Rome's rulers started to be chosen by the Praetorian Guard or by the army: the dynastic principle, which had largely worked well until then, was completely abandoned after the assassination of Severus Alexander (who was killed by his own troops). The legions, which were now mostly commanded by officers from the *equites* social class, began to kill and proclaim emperors according to their own interests. It was enough to offer a large donative to

Legionary from the Early Empire, preparing his full marching equipment. (*Photograph by Simone Ochsner (https://www.simoneochsner.com), copyright Legio XI CPF*)

the Praetorians or the legionaries for an ambitious man to obtain the throne, while any reduction in the pay of the soldiers could lead to him being deposed or killed. The weakness of the senatorial class and the continuous state of military emergency gave all political power to the military leaders, a state of affairs that temporarily ended only with the ascendancy of Diocletian.

Although for reasons of space we will focus specifically on the military changes that affected the Roman Army during the third century, it should be borne in mind that the whole Roman society changed in a dramatic way at this time. The Crisis of the Third Century was not only a political and military one. For instance, it also seriously reduced the demographic capabilities of the Empire due to the human losses caused by terrible epidemics. Economy and commerce were severely damaged too, while the traditional Roman way of life also had to change. New religions became gradually dominant and Roman society completely changed its internal organization. Traditionally, since the late Republic, Rome had developed a social structure that was based on three main classes: the aristocracy (senatorial families), *equites* (more or less a middle class) and the plebs (the lowest social group). The enormous difficulties of the

Legionaries from the Early Empire on the march. Note the use of a leather cover for the shields and the distribution of all the equipment on the back. (*Photograph by Simone Ochsner (https://www.simoneochsner.com), copyright Legio XI CPF*)

crisis reduced the number of social classes from three to two: the richest members of society, the aristocracy and the wealthiest *equites*, started to form a new group known as *honestiores*, while the poorer *equites* and the plebs became known as *humiliores*. In practice, the crisis had cancelled the middle class: the inhabitants of the Empire were now extremely rich or extremely poor. The most important *equites* had transformed themselves into a new ruling class, gradually making the traditional aristocracy of

Legionary officer ('*Praefectus Castrorum*') from the Early Empire, wearing muscle cuirass. (*Photograph by Simone Ochsner (https://www.simoneochsner.com), copyright Legio XI CPF*)

'*Aquilifer*' from the Early Empire, wearing a corselet of scale armour and carrying a ceremonial Imperial Eagle. (*Photograph by Simone Ochsner (https://www.simoneochsner.com), copyright Legio XI CPF*)

Centurion from the Early Empire, wearing a corselet of scale armour. (*Photograph by Simone Ochsner (https://www.simoneochsner.com), copyright Legio XI CPF*)

Centurion from the Early Empire, equipped with rectangular shield. (*Photograph by Simone Ochsner (https://www.simoneochsner.com), copyright Legio XI CPF*)

the senators more marginal. Obviously, this had very serious effects on the leadership of the army, as we shall see. To all this we should add another important social change, which happened before the beginning of the crisis. In 212, the *Constitutio Antoniana* of Emperor Caracalla had given Roman citizenship to all the inhabitants of the Empire. Until that moment, Rome's military organization had been based on the distinction between citizens and non-citizens. With the promulgation of the *Constitutio Antoniana*, however, the main social difference between legionaries and auxiliaries was completely cancelled. A provincial man now did not have to serve in the army as an auxiliary for many years in order to obtain Roman citizenship. Finally, it is important to note that the crisis had a deep impact on the urban centres of the Empire. Many historians have defined Roman civilization as an 'empire of the cities', because urban centres were the real core of the Roman state. The crisis changed all this: the large, open and rich cities of the previous decades were progressively abandoned due to the fear of foreign attack, while new, smaller cities started to be built, which were protected by defensive walls but were much more isolated and poor than before. The city of Rome was encircled by new walls, the construction of which began under Emperor Aurelian. From an economic point of view, the western provinces became increasingly poor due to them being exposed to foreign attacks; the eastern provinces, meanwhile, became increasingly rich thanks to their superior agricultural capabilities. The economic decline of the western territories affected the monetary system, which started to have a decreasing circulation of gold and silver coinage in the western provinces. This decisively altered the commercial balance of the Roman Empire, progressively destroying the economic integration of the Roman territories.

Chapter 2

The Roman Army of the Third Century

At the beginning of the Early Empire, the Roman Army was already a deadly military machine made up of professional soldiers. Following the Battle of Arausio (105 BC), the Roman military had been completely reformed by Gaius Marius, the great general and consul. Marius transformed the Roman Army into a force comprising professional fighters from the poorest social groups of the Republic. They served in Rome's armies for most of their life and were equipped by the state. The era of the Republican citizen-soldier was over: the new professional soldiers earned a living through their military career and were available to serve in every corner of Rome's expansive territories, with no time limits. Marius greatly improved the training and discipline of the Roman forces, giving each soldier heavy personal equipment that included several working tools, meaning each legionary (heavy infantryman) was able to operate as a combat engineer if needed, to build fortified encampments or destroy the fortifications of the enemy. The weight of equipment carried by the legionaries led to them being labelled as 'Marius' mules'. The ambitious general and consul strove to create a new *esprit de corps* within the legions: he was the first, for example, to give a distinctive standard and a peculiar denomination to each unit of the Roman Army. With the outbreak of the civil wars, the military reforms carried out by Marius were tested to the limit. Rome's opposing factions recruited armies comprising many thousands of professional legionaries, which destroyed each other in some of the bloodiest battles of Antiquity. No other army of the Mediterranean world could face the powerful new Roman legions on equal terms: they were better disciplined, trained, equipped and commanded than any of their opponents. When the civil wars came to an end in 30 BC, however, it became clear that the immense military machine that had been mobilized during the previous decades had to be greatly reduced in numbers. The finances of Rome could no longer sustain the cost of keeping in service an army with sixty legions and some 300,000 men. It should also be noted that during the latter years of the civil wars, several new legions had been formed with soldiers of inferior quality: sometimes every able-bodied man of a territory had been forced to enlist as a legionary, and the armies of various client kings were transformed into legions.

To create some form of 'peace establishment', Emperor Augustus had to drastically reduce the number of legions and carry out a rigid selection process in order to choose the best legionaries who would remain in service. The new army consisted of volunteers who wished to serve the Empire as professional soldiers for two decades; they were more loyal to the emperor than to any other institution. To ensure his soldiers' loyalty, Augustus created the *aerarium militare* (military treasure), which was specifically tasked with paying the legionaries with both precision and regularity. The emperor fully understood that the only way to avoid mutinies and rebellions was to pay the soldiers well (he increased their standard pay) and to offer them acceptable

Optio from the Early Empire, preparing his full marching equipment. (*Photograph by Simone Ochsner (https://www.simoneochsner.com), copyright Legio XI CPF*)

Vexillifer standard-bearer from the Early Empire. (*Photograph by Simone Ochsner (https://www.simoneochsner.com), copyright Legio XI CPF*)

Signifer standard-bearer from the Early Empire. (*Photograph by Simone Ochsner (https://www.simoneochsner.com), copyright Legio XI CPF*)

living conditions. Once a legionary completed his period of service, he was given a farm by the state as compensation for his twenty years of loyal service. As a result of these improved terms and conditions, the legions of the Roman Army began to consist of determined and disciplined men who believed that loyalty towards the emperor could guarantee them success in their personal life. To find the economic resources needed to create the *aerarium militare*, Augustus had to cut costs, so after 30 BC, the total number of legions was rapidly reduced from sixty to just twenty-eight. Only the best elements were retained in service. In order to avoid the outbreak of revolts, Augustus gave donatives and lands to most of the veterans who were discharged from the armies. During the period of the civil wars, the survival of a Roman soldier depended on the decisions of his commander, from whom he received his pay as well as a land property at the end of his career. Augustus, after his ascent to power, had to change all this and cancel the private dimension that had been acquired by Rome's military forces.

The main military unit of the Roman Army was the legion: a large heavy infantry formation, entirely composed of volunteers having Roman citizenship. Each legion comprised some 5,500 men, divided into one elite *Cohors Prima* and nine 'line' cohorts. The First Cohort, formed from the best soldiers of the legion, had 800 men, while each of the other nine cohorts numbered 480 legionaries. The *Cohors Prima* had five centuries of 160 men each, while the regular cohorts had six centuries of eighty legionaries each. Each *centuria* comprised ten smaller units, formed by seven soldiers plus a single NCO (known as a *decanus*) and two non-combatants; this *contubernium* was the smallest sub-unit existing within the Roman Army. In addition, each legion had its own cavalry component, with 120 cavalrymen divided into four *turmae*, each having thirty soldiers. Each *turma* was structured on three *decuriae*, the smallest cavalry sub-units, consisting of eight soldiers, a *decurio* (the equivalent of the infantry centurion) and an *optio* (acting as an NCO). Most of the legionaries came from Italy, but as time progressed the provincial elements became increasingly numerous. The system of ranks of the Roman legion was quite complex, being specifically designed to make such a large body of soldiers work properly. The overall commander of the legion was the *Legatus Legionis*, usually a senator appointed by the emperor, but over time many members of the *equites* social class took command of a legion. If a legion was garrisoned in a province where no other legions were located, its commander also acted as provincial governor; when performing these functions, he was commonly known as the *Legatus Augusti pro praetore*. The second officer in the chain of command was the Tribunus Laticlavius: quite often he was a young member of the Senate with no military experience, who was assigned to the position in order to learn from the Legate. The third in command was the *Praefectus Castrorum*, who was responsible for the building and defence of the legion's camp. He was a long-serving veteran from the *equites* social class, who had already finished his twenty years of service and had served as the most senior centurion in a legion. His functions included the training of soldiers and of new recruits. The group of senior officers was completed by five lower-ranking tribunes known as *Tribuni Augusticlavii*, who all came from the equestrian class and already had military experience. They mainly performed administrative functions and commanded two cohorts each.

The most senior centurion in a legion, the *Primus Pilus*, commanded the first century of the elite First Cohort. In battle, he had the honour of commanding the entire *Cohors Prima*. The other five centurions of the First Cohort were known as *Primi Ordines*, who were considered to be superior in rank to their equivalents of the other cohorts. The centurions commanding the first centuries of the nine line cohorts were called the *Pili Priores* and were superior in rank to the *Primi Ordines* of the First Cohort. Like the *Primus Pilus*, each of these officers took command of his entire cohort during battle. Generally speaking, all centurions came from the lower social classes and had become officers only through their personal military abilities, meaning these experienced veterans were the true backbone of the Roman Army. The commander of the four cavalry *turmae* was known as the *Tribunus Sexmentris*, while the decurion commanding each first *decuria* of horsemen had the honour of leading the entire *turma* in battle. The rank organization of the NCOs was particularly complicated, in order for them to perform a wide range of different functions. Each centurion directly appointed his adjutant and second in command, who was called an *Optio* and performed more or less the same command functions as the centurion, but during battle he had the important responsibility of remaining at the back of the century in order to keep the formation closed-up and to discourage deserters from abandoning the field. Each *Optio* had an adjutant known as a *Tesserarius*, who was third in the century's chain of command; he was keeper of the watchword of his unit and performed several important administrative functions. The system of NCOs was completed by the ten *Decani*, each of whom commanded one of the *contubernia* that made up a century.

Common legionaries could perform a series of special duties which had a specific designation and corresponded to higher pay. The most important of these was that of *Aquilifer*. The *Aquila* was the legion's standard and represented the valour of the whole unit; losing it was the greatest dishonour that a Roman legion could endure. As a result of this symbolic importance, its bearer was chosen from the most experienced veteran soldiers. Each cohort had its own *Vexillum* (standard) that was carried by a special bearer known as a *Vexillifer*, while each century had its own *Signum* that was carried by a *Signifer*, which consisted of a spear shaft decorated with medallions

Signifer standard-bearer from the Early Empire. (*Photograph by Simone Ochsner (https://www.simoneochsner.com), copyright Legio XI CPF*)

Signifer standard-bearer wearing corselet of chainmail. (*Photograph by Simone Ochsner (https://www.simoneochsner.com), copyright Legio XI CPF*)

and topped with an open hand to signify loyalty towards the state. In addition to the above, the *Cohors Prima* had the privilege to carry a special image of the emperor, which was given to an elite bearer known as the *Imaginifer*. Finally, each century had its own military music, consisting of a *Cornicen* or *Bucinator* and a *Tubicen*. All these transmitted orders by playing their instruments, and usually acted in close combination with the *Signifer*, who was a rallying point for the legionaries. The *Cornicen* played a horn, the *Bucinator* had a particular kind of horn known as a *buccina* and the *Tubicen* played a long trumpet. Each legion deployed its own artillery in addition to its own cavalry: each cohort employed a stone-throwing *ballista* and each century had a bolt-firing *scorpio*. A single Roman legion could deploy an impressive artillery detachment, with ten heavy *ballistae* and fifty-nine light *scorpiones*. Under the military reforms carried out by Gaius Marius, the Roman legions adopted a new structure that had no light infantry but only *cohortes* of heavy infantry. Consequently, the Romans had to rely on foreign soldiers during the late Republican period in order to deploy sizeable contingents of light troops. The same happened with the cavalry, which had never been an important component of the Roman Army.

When Augustus reformed the structure of Rome's military, he decided that the foreign contingents of light infantry and cavalry should become a fixed component of the Roman Army. Until that moment,

Signifer standard-bearer wearing the skin of a bear. (*Photograph by Simone Ochsner (https://www.simoneochsner.com), copyright Legio XI CPF*)

these light and mounted contingents had been recruited on a temporary basis and had always been disbanded at the end of a military campaign. They did not have a standardized organization and could be commanded by their own officers, who were mercenaries paid by the Roman Republic. Augustus changed all this by transforming the provisional corps of light infantry and cavalry into a permanent component of the new Roman Army: the *auxilia*, or auxiliaries. These were commanded by Roman officers and received regular pay from the imperial government; as a result, they lost their mercenary status and became a regular military institution. The new auxiliaries, like their predecessors, were recruited on a local basis from the provincial inhabitants of the Empire and thus provided contingents with different tactical features, coming from every corner of the Roman world. Differently from the legionaries, they were not Roman citizens, but after twenty-five years of military service they were granted Roman citizenship and a farm located in their home province. As a result, thanks to the new system introduced by Augustus, serving in the *auxilia* units became the best way for a provincial to become a Roman citizen and improve his social position. Thanks to the sacrifices of his father, the son of a provincial auxiliary could enlist in the Roman Army as a legionary. A new hereditary class of provincial soldiers was soon formed, with men who were strongly linked to their home provinces but at the same time were extremely loyal to Rome thanks to the opportunity of social improvement that they had received. As time progressed, the auxiliary units increased their level of professionalism and started to be sent on campaign across the Roman territories. With the expansion and consolidation of the Empire, new military traditions were gradually absorbed into the Roman Army thanks to the *auxilia* system, such as archers recruited from the eastern provinces or camel troops.

Basically, there were three different kinds of auxiliary units: *cohortes* of all infantry, *alae* of all cavalry and *cohortes equitate* (infantry cohorts with an attached cavalry contingent). The auxiliary cohorts had the same internal organization and structure as the legionary ones, albeit acting as light infantry units. The *alae* were the true cavalry of the Roman Army, as the few cavalrymen of the legions performed only auxiliary duties for the heavy infantry (escorting officers, transmitting orders or scouting). The *alae* could be either of heavy or light cavalry, but were all characterized by the quality of their specific training. Some of them were even mounted on dromedary camels. The Romans first encountered camel-mounted troops when they fought against the Seleucid Antiochus III in 190 BC, but it was only during the second century AD that a Roman camel corps was formed. In AD 106, Trajan added to the Empire the new province of Arabia Petraea and created the *Ala I Ulpia Dromedariorum Palmyrenorum*, which comprised camel warriors from the city of Palmyra. In the following decades, the Romans formed various small auxiliary units of *dromedarii*, which proved very

Legionary standard-bearer from the Early Empire, wearing the skin of a wolf. (*Photograph by Simone Ochsner (https://www.simoneochsner.com), copyright Legio XI CPF*)

Legionary standard-bearer equipped with chainmail and metal greaves. Note the rich decorations painted on the round shield. (*Photograph by Simone Ochsner (https://www.simoneochsner.com), copyright Legio XI CPF*)

Cornicen musician from the Early Empire. (*Photograph by Simone Ochsner (https://www.simoneochsner.com), copyright Legio XI CPF*)

effective in patrolling the desert frontiers of the Empire. Their tasks included escorting convoys, defending important routes of communication, scouting in the desert, escorting couriers and battling against desert bandits. They were organized in small squads, having special bases in the desert that provided them with food and water. The *cohortes equitate* – created by Julius Caesar – were a mixture of foot and mounted auxiliaries, a smaller replica of the larger legions. Thanks to the presence of some horsemen, they could act as small autonomous armies for service on the frontiers. Each of these three kinds of auxiliary units could have two different establishments, with 500 or 1,000 men. A *cohors quingenaria* had six centuries of eighty soldiers, with a *Praefectus* as commander, while a *cohors miliaria* had ten centuries of eighty soldiers, led by a *Tribunus Militum*. The *ala quingenaria* had sixteen *turmae* of thirty men and the *ala miliaria* had twenty-four turmae of thirty men, both units having a *Praefectus* as commander. A *cohors equitata quingenaria* had six centuries of eighty infantrymen and four *turmae* of thirty cavalry, being commanded by a Praefectus, and a *cohors equitata miliaria* had ten centuries of eighty infantry and eight *turmae* of thirty cavalrymen, being commanded by a *Tribunus Militum*.

During military campaigns, the various auxiliary units were attached to the legions, thus coming under direct control of the latter's commanders. Most of the *auxilia* units were stationed on the vast borders of the Empire, together with the legions. The frontiers were the only points from which enemy attacks could come, so this defensive disposition of the Roman forces was quite logical. It should be pointed out, however, that in the event of internal disturbances, very few men were available to face threats coming from inside the Empire. There was no central military reserve and it was practically impossible to take a full legion from its frontier garrison without creating a major gap in the border defences. The only possible solution to this serious problem was to take detachments from various legions and assemble them together in order to form temporary task forces known as *vexillationes*. As soon as the internal emergency was resolved, the temporary *vexillatio* was dissolved and its detachments were sent back to their parent legions. Usually, a *vexillatio* was a mixed force of legionaries and auxiliaries, comprising 1,000 infantrymen and 500 mounted troops: two infantry cohorts taken from a legion and a single *ala quingenaria* of mounted auxiliaries.

The military situation in Italy and the city of Rome was quite different from that of the provinces. The capital was the heart of the Empire, where the emperor and the entire central administration worked. According to the reforms of Sulla and to ancient Roman tradition, no military units could enter the Italian peninsula without representing a menace for the institutions of Rome: the territory of Italy was considered sacred by the Romans, who called it the *pomerium*. This corresponded more or less to modern Italy, with the Rubicon River in the north and the Strait

Cornicen musician from the Early Empire, wearing the skin of a wolf. (*Photograph by Simone Ochsner (https://www.simoneochsner.com), copyright Legio XI CPF*)

Legionary officer from the Early Empire, equipped with Hallaton helmet and muscle cuirass. (*Photograph by Simone Ochsner (https://www.simoneochsner.com), copyright Legio XI CPF*)

of Messana in the south being its natural borders. As a result of this situation, Italy could not be garrisoned by legions or by auxiliary units like all the other provinces. Obviously, however, the emperor and the city of Rome needed some military forces for protection (especially against internal threats). Augustus resolved this issue by creating the Praetorian Guard. Formally the personal guard of the emperors, this

Centurion from the Early Empire, wearing the helmet with transverse crest that was distinctive of his rank. (*Photograph by Simone Ochsner (https://www.simoneochsner.com), copyright Legio XI CPF*)

unit also acted as the garrison of the Italian peninsula. Initially, it was formed from the best soldiers from the legions, and thus represented the elite of the Roman Army. Created by Augustus and reorganized by his successor, Tiberius, the Praetorian Guard was structured on nine infantry cohorts. Under the first emperor, three cohorts were stationed in Rome and the remaining six in other important cities of Italy. However, during the reign of Tiberius, all the Praetorians were garrisoned in Rome, in the barracks known as the *Castra Praetoria*. The whole Praetorian Guard was commanded by a *Praefectus Praetorio*, who came from the ranks of the *equites*. Each cohort was commanded by a *Tribunus Militum* and had six centuries of eighty soldiers. The original number of nine cohorts was later changed on several occasions: around AD 47 they were increased to twelve, and in AD 69 they became sixteen (under Vitellius, who also augmented their internal establishment from 500 to 1,000 men). Vespasian reduced the number of cohorts again to nine (with 500 men each), while Domitian increased it to ten by setting the definitive establishment that would not change until the final dissolution of the corps. Initially, all the Praetorians were ex-legionaries from Italy, but over time there was an increasing presence of provincials in their ranks. The soldiers of the Praetorian Guard had many privileges compared with the normal legionaries: their period of service lasted for sixteen years instead of twenty and their pay was significantly higher than that of regular soldiers. Trajan added a cavalry component to the Praetorian Guard, whose members were known as *Equites singulares Augusti*. These mounted Praetorians escorted the emperor whenever he left Rome for a military campaign or for a journey in the provinces. The *Equites singulares Augusti* were organized as a normal *ala miliaria*, with twenty-four *turmae* of thirty men and a *Tribunus Militum* as commander. All members of this unit were selected from the best horsemen of the auxiliary *alae* and were granted Roman citizenship upon enlisting into the Praetorian Guard.

The military reforms of Augustus that affected the garrison of Italy were not limited to the creation of the Praetorian Guard, as he also formed another two corps that were to be stationed in Rome: the *Urbaniciani* and the *Vigiles*. The first were a sort of police, whose main function was to keep order in the immense city of Rome. If the Praetorians were the protectors of the emperor, the *Urbaniciani* were the protectors of Rome. While the Praetorian Guard was under the direct control of the emperor, the *Urbaniciani* were the only Roman military unit still controlled by the Senate: their commander, the *Praefectus Urbi*, was in fact a senator. Initially, the *Urbaniciani* consisted of three cohorts, numbered progressively after the nine of the Praetorians. Each cohort was commanded by a *Tribunus Militum* and had six centuries of eighty soldiers. The original three cohorts were later expanded to seven during the reign of Claudius, while Vitellius increased the size of each cohort to 1,000 men. Differently from the Praetorians, the *Urbaniciani* continued to be recruited from Italy during their entire existence. The other corps created by Augustus and stationed in Rome, the *Vigiles*, were a sort of urban guards/firemen. Their main functions were patrolling the streets during the night and fighting fires (which could be destructive for a city of the time, as the Great Fire of Rome showed in AD 64). Their commander was known as the *Praefectus Vigilum* and came from the class of the *equites*. They were organized in seven *cohortes miliariae* for a total of 7,000 men. Most of their members were *liberti*, freed slaves of low social condition. Each cohort of the Vigiles was under the orders of a *Tribunus Militum* and was peculiarly divided into just seven centuries. At the beginning of the Empire, the city of Rome was divided into fourteen regions, so each cohort of firemen was responsible for the security of two *regiones*.

Gallienus, who ruled the Roman Empire from 260–268, is commonly considered to be the first Roman monarch who understood that the military organization created by Augustus had to be completely changed in order to have a more effective defence of the Empire. The traditional structure consisting of legions and *auxilia* units was too static to deal with the massive migrations of Germanic peoples or the large-scale invasions of the Sassanids. In addition, civil unrest and internal threats had to be taken into consideration. Until the beginning of the Crisis of the Third Century, the Roman Army had only been employed to fight on the frontiers, but the internal secessions had clearly shown the need to use the armed forces to perform police functions. One of the consequences of the crisis was the birth of a new phenomenon of internal unrest, a type of brigandage that saw the formation of large bands of insurgents/criminals (commonly known as *Bagaudes*). As a consequence, the army had to be ready to maintain public order in addition to serving on the frontiers. The *Constitutio Antoniana* of Caracalla had practically ended any possible future for the *auxilia* system. Gallienus tried to resolve all these problems, starting with the defence of the frontiers. The idea of a static line of defence had been totally surpassed by events: the Empire now needed central reserves of highly mobile troops which could move very rapidly to meet various threats in several points of the *limes*. The only units that could deploy such a level of mobility were the cavalry. The legionary or auxiliary infantry garrisoned on the borders were unable to fight successfully against simultaneous threats. Initially, Gallienus created four small cavalry reserves, located in strategic places of the Empire: Mediolanum in northern Italy, Sirmium in Serbia, Poetovio in Slovenia and Lychnidos in Macedonia. Mediolanum and Sirmium eventually became the most important locations, with two large strategic reserves of cavalry stationed there. The troops in Mediolanum were ready to counter any emergency menacing Italy, while the troops in Sirmium could act along the long Danubian *limes*.

Legionary cavalryman wearing helmet with facial mask. (*Photograph and copyright by Robert Vermaat / Ala I Batavorum*)

Legionary cavalryman wearing parade equipment. (*Photograph and copyright by Robert Vermaat / Ala I Batavorum*)

Auxiliary cavalryman from the Rhine frontier, equipped with parade helmet of the 'Nijmegen-Kops Plateau' model. (*Photograph by Simone Ochsner (https://www.simoneochsner.com), copyright Legio XI CPF*)

As a result of Gallienus' cavalry reforms, mounted units swiftly became much more important than before. As we have seen, the Romans had never considered cavalry a very important part of their military, but the *Constitutio Antoniana* had transformed all provincials into Roman citizens so new cavalry units could now be easily recruited in all the provinces of the Empire. The collapse of the *auxilia* system and the contemporary expansion of the cavalry led to the birth of several new categories of mounted troops, which made up the new strategic reserves and had some innovative characteristics. Before analysing these new troop types, however, it is important to consider the other military reforms produced by Gallienus. These mostly affected the leadership of the army, with the senatorial class being formally excluded from any form of military service. The aristocracy, which for centuries had controlled the entire Roman military machine, was in such a state of decay that the emperor could assign command of the whole military apparatus to the richest members of the *equites*. In effect, Gallienus had not done anything revolutionary; he merely transformed an informal practice into a formal one. Since the time of Marcus Aurelius, senators had gradually and voluntarily abandoned their command of the legions, preferring to dedicate themselves to much more lucrative activities like trading. The traditional ethos of the Roman aristocrats had disappeared alongside the ancestral values of Roman society. Fortunately for Rome, however, this important heritage was then accepted by the equestrians of the new *honestiores* social class, who were ready to take the place of the senators in the ranks of the army. The *Legatus Legionis* was thus replaced by the *Praefectus Legionis*, who was a member of the *equites*. The general expansion of cavalry was not limited to the creation of new separate mounted units, but involved the transformation of legionary cavalry. Since the time of Augustus, the cavalry of each individual legion had consisted of just 120 horsemen divided into four *turmae*, with thirty soldiers each. But with the reforms of Gallienus, each legion's cavalry was expanded to the establishment of an *ala miliaria* (twenty-four *turmae* of thirty men each, for a total of 720 horsemen).

The new types of cavalry units created by Gallienus were the direct heirs of the former mounted *auxilia*. Very little is known about their internal organization, but what we do know is that they all assumed the general denomination of *vexillationes*. Indeed, from the reign of Gallienus, this term was no longer used to identify detachments taken from the legions; the *vexillatio* was now the standard cavalry unit, which could comprise various different kinds of *equites*. What follows is a brief analysis of the new categories of *equites* created by Gallienus.

Equites Dalmatae: like all the new cavalry *vexillationes* created by Gallienus, these originated as detachments of existing mounted units (both legionary cavalry and *auxilia*) that were permanently separated from their parent corps and thus constituted

Legionary from the early third century wearing *lorica segmentata*. (*Photograph and copyright by Paul Elliott*)

Legionary from the early third century equipped with oval shield. (*Photograph and copyright by Paul Elliott*)

Legionary from the early third century resting during a march. (*Photograph and copyright by Paul Elliott*)

new independent units. It is highly probable that this process was caused by the military emergency experienced by the Roman Empire during the reign of Gallienus. With the division of the Empire caused by the two large secessions of Gaul and Palmyra, only the central territories of Italy and the Balkans remained under the direct control of Gallienus. As a result, many detachments originating from Gaul or the eastern part of the Empire that were temporarily stationed in the Balkans found themselves isolated from their parent units. Gallienus transformed these into independent units, and the term *vexillationes* gradually started to denote independent mounted units. The *Equites Dalmatae* were probably the first new category of cavalry to be formed, since they were quite numerous: the *Notitia Dignitatum*, an important primary source produced several decades after Gallienus' cavalry reforms, lists forty-eight units of this kind. The term *Dalmatae* had no specific ethnic meaning, it simply being used to indicate the geographical provenance of these units: they were originally stationed in the region of Dalmatia, but were composed of soldiers coming from every corner of the Empire. It would be totally wrong to think that these units were entirely raised from inhabitants of the Balkan provinces, because this would not explain the survival of such units after the end of the Crisis of the Third Century military emergency. It is true that the Balkan provinces were the military core of Gallienus' truncated Empire, but they alone could not deploy such a large number of cavalry units. With the progress of time, the original detachments stationed in the Balkan provinces under Gallienus started to be located in other areas of the Empire, but they apparently retained their original denomination of *Dalmatae* to show that they had formerly been part of Gallienus' forces. During the terrible years of the secessions, the emperor had to face a severe military crisis and was obliged to use all the military resources that were available in the territories under his control. Apparently, the *Equites Dalmatae* were permanently assigned to the garrisons of frontier provinces as highly mobile cavalry forces that could deal with enemy incursions.

Equites Illyriciani: these had the same origins as the *Equites Dalmatae*, but were initially stationed in a different geographical area. While the *Equites Dalmatae* were detachments garrisoned in Dalmatia, the *Equites Illyriciani* came from nearby Illyricum. Dalmatia and Illyricum were the heart of the Danubian provinces in the Balkans, where many *vexillationes* from other areas of the Empire were located. When Gallienus had to build a new army from his truncated territories, he used cavalry detachments as the backbone of his new military forces and transformed them into independent units. The *Notitia Dignitatum* lists twenty-three units of *Equites Illyriciani*, garrisoned in various border provinces of the Empire. Like the *Equites Dalmatae*, they started to be used as frontier cavalry during the reign of Diocletian.

Equites stablesiani: their history and development was exactly the same as the *Equites Dalmatae* and *Equites Illyriciani*, the only difference being their regional provenance. The *Equites stablesiani* were created by converting the *vexillationes* garrisoned in northern Italy into independent cavalry units. As we have already seen, Italy and the Balkan provinces were the only territories that remained under the direct control of Gallienus. Looking at the original four strategic cavalry reserves created by the emperor, they were located in northern Italy and the Balkans: Mediolanum in northern Italy, Sirmium in Serbia, Poetovio in Slovenia and Lychnidos in Macedonia. It is thus reasonable to suppose that the *Equites stablesiani* were based in Mediolanum, the *Equites Dalmatae* in Sirmium and Poetovio, and the *Equites Illyriciani* in Lychnidos. The term *stablesiani* comes from the particular title of *stabulensis* that was given by Gallienus to Aureolus, who was the first commander of the strategic cavalry reserve based in Mediolanum. *Stabulensis* has a merely honorific meaning, designating Aureolus as the senior officer in charge of the imperial stables. The *Notitia Dignitatum* lists a total of sixteen units of *Equites stablesiani*. Like the similar *Dalmatae* and *Illyriciani* units, the *Equites stablesiani* also later became frontier cavalry.

Legionary from the early third century wearing *lorica segmentata*. (*Photograph and copyright by Paul Elliott*)

Legionary from the early third century wearing segmented armour. (*Photograph and copyright by Paul Elliott*)

Legionary from the early third century equipped with hood of chainmail. (*Photograph and copyright by Paul Elliott*)

Equites promoti: as a result of Gallienus augmenting the numerical composition of the legionary cavalry to the establishment of an *ala miliaria* (twenty-four *turmae* of thirty men each, for 720 horsemen), the cavalry of each legion now had the necessary numbers to act in more independently. During the military emergency caused by the invasions of the Germanic tribes and by the internal secessions, Gallienus transformed several formations of legionary cavalry into independent units: that is the reason why these were known as 'promoted'. It is reasonable to suppose that the *Equites promoti* retained their original internal organization, at least initially. The *Notitia Dignitatum* mentions thirty-one units of *Equites promoti*. Like the other new cavalry units created by Gallienus, they were progressively garrisoned in all the provinces of the Empire and used as frontier cavalry.

Equites indigenae: during the early centuries of the Empire, the great majority of the Roman Army's cavalry had been composed of provincial auxiliaries, but when the *Constitutio Antoniana* gave Roman citizenship to all provincials, the status of these auxiliary cavalrymen suddenly changed. Most of the cavalry units formed by raising local elements in the various provinces started to be defined as *indigenae* (indigenous), abandoning their auxiliary status but retaining their local nature. In total, the Notitia Dignitatum names thirty-eight units of 'native cavalry', stationed in every corner of the Empire (mostly along the frontiers). Every province had its own units of native cavalry, who were strongly influenced by the local military traditions of their territories.

Equites Mauri, Equites scutarii, Equites sagittarii: these three categories of cavalry were not directly related to the reforms of Gallienus, but were part of the new cavalry organization that emerged after the changes of the late third century. Moorish (i.e. Numidians and Mauri) light horsemen had long been a fundamental component of the Roman Army's auxiliary cavalry. Their role was particularly important during the campaigns fought against the Parthians or the Sassanids in the East, because they were excellent skirmishers armed with javelins and mounted on fast horses. The Moorish light skirmishers were the perfect Roman tactical response to the large numbers of horse archers deployed by the Parthians and Sassanids. They remained a military elite after the Crisis of the Third Century, so although the *auxilia* system had long since disappeared, there are ten Moorish light cavalry units listed in the *Notitia Dignitatum*. These were garrisoned in different areas of the Empire and formed a distinct category of troops, separated from that of the *Equites indigenae* even in the provinces of North Africa, which were the home of the Moorish tribes. The existence of a specific denomination for Moorish units confirms the high esteem in which they were held by the Romans, especially during a period that was characterized by a strong need for new light units (both mounted and on foot). The *Notitia Dignitatum* also includes eight units of Moorish infantry, a clear signal that the *Mauri* fighters were also highly appreciated as foot skirmishers. The *Equites scutarii* were a category of mounted troops about which very little is known. The term *scutarii* does not mean that they were all equipped with shields, because during the last centuries of the Empire it was used to indicate guard units. It is thus plausible to presume that the seventeen units of *Equites scutarii* named in the *Notitia Dignitatum* acted as mounted bodyguards, although we have no clear idea of their actual functions. It is possible that they were used as the mounted escort of high-ranking officers. The difficult wars fought against the Sassanids convinced the Romans to include some horse archers in their armies. These new units, commonly known as *Equites sagittarii*, were formed in all the provinces of the Empire; the Notitia Dignitatum lists fifty-one of them.

The Roman military reforms of the third century were not limited just to the cavalry; the legionary infantry was deeply affected by them too, especially with regard to light troops. As we have seen, the Roman Army had employed provincial soldiers for the majority of its light infantry units. However, it should not be forgotten that the legions had always comprised a certain number of lightly equipped infantrymen. These were known as *antesignani*, and there is trace of them since the times of Julius Caesar. The *antesignani* can be defined as elite light infantrymen who made up the vanguard of each legion, frequently being used for special missions and carrying lighter equipment than the normal heavy infantry legionaries. They were not part of the heavy infantry formations and used small oval shields instead of the bigger ones carried by standard legionaries. During marches, their main functions were to explore the territory through which they were moving and to protect the flanks of the legion. In battle, meanwhile, they countered the missile troops of the enemy and supported cavalry attacks thanks to their superior mobility. It is important to note that, despite their important functions, the *antesignani* did not exist as a separate corps within the legions. They were not a standing unit, but were drawn from the ranks of the normal cohorts when tactical considerations demanded it. Their assignments, albeit very important, were always temporary. The *antesignani* were usually drawn from the youngest and fittest legionaries of each cohort in order to perform special missions that needed a high level of mobility.

During the third century, in particular under the reign of Septimius Severus, the light infantry component of the legions started to be increasingly important and acquired a stable organization. Apparently, the first legions to have a fixed number of light infantrymen, who performed as skirmishers on a permanent basis, were the three new ones created by Septimius Severus for his Parthian campaign. Since their first encounters with the Parthian mounted archers, the Romans had suffered from a lack of missile troops in their heavy infantry formations. It seems that Septimius

Draconarius (left) and officer (right) from the central decades of the third century, serving on the Germanic frontier. (*Photograph and copyright by Conrad Lunar*)

Officer from the mid-third century, wearing scale cuirass. (*Photograph and copyright by Conrad Lunar*)

Officer (left) and legionary (right) from the mid-third century, serving in a fort of the Germanic frontier. (*Photograph and copyright by Varianus/Numerus Brittonum*)

Severus tried to resolve this problem by including a fixed number of light infantrymen in his three newly raised elite legions. According to the new structure introduced for these formations, and later adopted by all the legions, each cohort was to include two centuries of light infantry and four of heavy infantry (the elite *Cohors Prima* had two centuries of light infantry and three of heavy infantry). The transformation of two centuries from heavy to light infantry clearly shows the increasing need for missile troops. The new Roman light infantrymen were known as *lanciarii*, from the name of their main weapon (the *lancea*). Quite frequently, on special occasions,

Officer from the mid-third century with Niedermörmter helmet. (*Photograph and copyright by Varianus/Numerus Brittonum*)

Officer from the mid-third century wearing scale corselet. (*Photograph and copyright by Varianus/Numerus Brittonum*)

Centurion from the central decades of the III century with Niederbieber helmet. (*Photograph and copyright by Varianus/Numerus Brittonum*)

all the centuries of *lanciarii* included in a legion could be brigaded together and detached in order to act independently. This means that, like the *antesignani*, they were also considered as elite troops. Over time, however, the *lanciarii* gradually transformed themselves into elite guard units and partly abandoned their original light infantry nature. By the time of Diocletian, they had completed this transformation and started to be grouped into large independent units. The *Notitia Dignitatum* records that the Roman Army included nine legions entirely formed of *lanciarii*.

Finally, we should note that during the turbulent decades of the third century, the Roman generals started to make increasing use of irregular units entirely composed of barbarian soldiers. Germanic warriors were generally very happy to serve in the Roman Army as mercenaries, but were always reluctant to integrate themselves into regular military units. As a result of this situation, the Romans created two new kinds of military units that were entirely composed of barbarian fighters: the infantry *numerus* and the cavalry *cuneus*. During the following centuries, these units were also opened to Romans and started to receive a more formal internal structure. During the third and early fourth centuries, however, they remained corps of barbarians of a semi-regular nature. Apparently, neither the *numerus* nor the *cuneus* had any standard structure to speak of; the groups of German warriors could be very numerous or quite small, and usually had no internal subdivisions. What is known is that the *numeri* and *cunei* fought in their own style and often under their own chieftains, who contracted their services to the Roman military authorities. Initially, these units of hired barbarians were mostly employed along the borders, as a supplement to the units of *auxilia* guarding the frontier. Over time, they became increasingly numerous and important, transforming themselves into a significant portion of the Roman military.

Chapter 3

The Wars of the Middle Empire

The death of Marcus Aurelius in AD 180 heralded a phase of political instability for the Roman Empire, as his son and successor, Commodus, soon proved to be mentally insane and unfit to rule the Roman state. The new emperor ordered the execution of several aristocrats who had been loyal to his predecessor, greatly reduced the influence of the Senate, spent large sums of money in organizing magnificent gladiatorial games in the Colosseum and did nothing to stabilize the military situation on the Germanic *limes*, where the Marcomanni and the Quadi had recently been defeated by Marcus Aurelius. In 193, three Roman nobles, fearing that they could be the next targets of Commodus' paranoia, plotted the assassination of the emperor, who was strangled in the arena while fighting as a gladiator. Pertinax, the experienced Proconsul of Africa, was chosen as the new ruler of Rome because of the positive relations he had with many members of the Senate. The emperor, however, soon had to face a series of problems, being accused of having ordered Commodus' assassination and having to resolve the dramatic financial crisis that was affecting the Roman Empire. When Pertinax refused to pay a large *donativum*, or gift, to the Praetorians, they rebelled against him, wishing to keep the privileges that had been assigned to them by Commodus. After reigning for just three months, Pertinax was assassinated by the Praetorians, who subsequently auctioned off the title of emperor to the highest bidder. Didius Julianus, who had already succeeded Pertinax as Proconsul of Africa, was thus able to win the favour of the Praetorian Guard and became emperor. Soon after his ascendancy, however, two important military leaders rose up in revolt against him: the governor of Syria, Pescennius Niger, and the governor of Pannonia, Septimius Severus. The latter marched with his legions on Rome and ordered the execution of Didius Julianus, who had been emperor for less than two months. Both Septimius Severus and Pescennius Niger were then proclaimed emperors by their troops; 193 later became known as the Year of the Five Emperors, since it saw five different rulers on the imperial throne: Commodus, Pertinax, Didius Julianus, Pescennius Niger and Septimius Severus. Pescennius Niger and Septimius Severus fought each other in a bloody civil war that lasted until 194 and ended with the victory of Septimius Severus.

Centurion from the mid-third century wearing scale cuirass and metal greaves. (*Photograph and copyright by Varianus/Numerus Brittonum*)

Despite overcoming his rival, Septimius Severus had to continue campaigning until 197 in order to pacify the Empire because one of his strongest political allies – Clodius Albinus – revolted against him. In 197, Septimius Severus decisively defeated Clodius Albinus at the Battle of Lugdunum and thus remained the sole ruler of Rome. He was a very capable and ambitious military leader, who wanted to expand Roman territories both in Britain and the Middle East. In 197, he went to the eastern provinces of his empire at the head of a large army, with the intention of invading Parthian territory. During his preparations for this campaign, he raised three new legions – *I Parthica*, *II Parthica* and *III Parthica* – that had an innovative character since they comprised two centuries of light infantry in each of their cohorts. The emperor, realizing that his army needed elite light infantry forces to face the Parthians on equal terms, equipped some of his newly raised legionaries as javelineers. After the end of the Parthian campaign, the *II Parthica* was transferred to Italy, the first Roman legion to be garrisoned in the Italian peninsula since the reign of Augustus. Septimius Severus made this move because he disliked the Praetorians and wanted to replace them with his own loyal veterans of the *II Parthica*, who were specifically tasked with preventing internal rebellions while acting as a strategic reserve that could be sent to any corner of the Empire in case of need. In practice, it was the first form of *comitatus*

Officer from the years of the Tetrarchy, wearing ridge helmet with crest and muscle cuirass. (*Photograph by Rachele Lori, copyright by Protectores Domini Nostri and Legio Palatina Ioviani Seniores*)

Officer from the years of the Tetrarchy, wearing ridge helmet with metal crest and muscle cuirass. (*Photograph by Rachele Lori, copyright by Protectores Domini Nostri*)

praesentalis (literally 'the army in the imperial presence'; see Chapter 4). The Parthian campaign of Septimius Severus gained some notable successes: his troops raided the Parthian imperial city of Ctesiphon and annexed a portion of Mesopotamia to the Roman Empire. Despite twice besieging Hatra, however, the emperor was unable to take the important Parthian stronghold. Septimius Severus was also active in other areas of the Empire. Having been born in Africa, he always paid special attention to the Roman territories located along the southern coastline of the Mediterranean. During 202, for example, he significantly expanded the Roman presence in Africa at the expense of the warlike Garamantes and refortified the *limes* that separated the Roman provinces from the nomadic tribes living in the Sahara.

In 208, Septimius Severus invaded Caledonia – present-day Scotland – in order to secure the Roman presence in Britannia. For more than a century, the Celtic tribes of Caledonia – which became known as Picts – had conducted raids and incursions into Roman territories in northern Britannia. Even the construction of Hadrian's Wall and the lesser-known Antonine Wall had not been enough to stop their activities. The portion of Caledonia between Hadrian's Wall in the south and the Antonine Wall in the north, which largely corresponded to the Lowlands of Scotland, had been temporarily controlled by the Romans. By 208, however, the Antonine Wall needed urgent repairs and the Roman Army exerted its direct control only south of Hadrian's Wall. Septimius Severus arrived in Britannia at the head of 40,000 soldiers, with whom he rapidly marched north. He initiated a massive rebuilding project of Hadrian's Wall, with the whole wall completed in stone, the western portion of the fortification having previously been made of just turf and timber. The Romans reoccupied the whole area south of the Antonine Wall, which had been lost to the Celtic tribesmen over recent decades, and completely repaired the northern barrier. In 209, Septimius Severus launched an invasion of the heartland of the Caledonians, the wild Highlands. Only the great Roman general Agricola, a century before him, had attempted anything similar. The Caledonians responded to the invasion with guerrilla tactics, which caused severe losses to the Romans. Septimius Severus employed very harsh methods against the Caledonians, devastating the lands of all the tribes that did not submit and killing thousands of civilians. In 210, when it seemed that the whole of Caledonia could be conquered by the Romans, Septimius Severus fell ill and was forced to go back to York to recover. However, he died in York in February 211. The emperor was succeeded by his son, Caracalla, who had participated with distinction in the war against the Caledonians. The new monarch, being obliged to consolidate his power in Rome, had no choice but to suspend the invasion and leave Britannia with most of his troops. The military events of 208–210 were the last Roman attempt to invade Caledonia: they greatly weakened the Celtic

Officer from the years of the Tetrarchy, with *pileus pannonicus* made of fur and muscle cuirass. (*Photograph by Rachele Lori, copyright by Protectores Domini Nostri*)

Officers from the years of the Tetrarchy. Note the complexity of the ornamental embroidering applied on the tunics. (*Photograph by Rachele Lori, copyright by Protectores Domini Nostri*)

tribes of Scotland and secured the survival of Britannia for many years. Nevertheless, soon after Caracalla's departure, the Antonine Wall was abandoned again and the Lowlands were retaken by the Caledonians.

Caracalla, like his father, spent large sums of money to improve the combat power of the Roman Army and continued to nurture the ambition of conquering

Emperor from the years of the Tetrarchy. The helmet is decorated with gems and with peacock feathers. (*Photograph by Rachele Lori, copyright by Protectores Domini Nostri*)

the Parthian Empire. Differently from Septimius Severus, who believed that the Parthian mounted archers could be defeated by well-trained light infantry, Caracalla was convinced that the only way to defeat the excellent Parthian cavalry was to re-equip the Roman legionaries as phalangites in imitation of Alexander the Great's Macedonian armies. Caracalla levied a new military unit from the Roman province of Macedonia, its 16,000 men being trained and equipped as phalangites. The new corps – corresponding to three legions – was only short-lived and did not play a significant role during the subsequent Parthian campaigns due to the failure of Caracalla's anachronistic tactical experiments. Between 215 and 217, the emperor campaigned with significant military forces in the Middle East, devastating a large portion of the Parthian Empire, aided by the internal divisions that damaged his enemies significantly. In 217, however, Caracalla was killed in a plot organized by the commander of the Praetorian Guard, having been unable to consolidate his new conquests in Mesopotamia.

Between 212 and 217, Caracalla had faced a new military threat on the northern borders of the Empire: the Alamanni. These were a confederation of Germanic tribes, whose name translates as 'all men'. The Alamanni attacked the Roman frontier in its most exposed area, the *Agri Decumates* (literally 'Decumatian Fields'), part of present-day south-western Germany that linked up the *limes* of the Rhine with that of the Danube. The Decumatian Fields were the only place in continental Europe where the border of the Roman Empire was not marked by the Rhine or the Danube, meaning they were particularly exposed to foreign invasions and had been heavily fortified by the Romans. By conquering the Decumatian Fields, an enemy of Rome could easily outflank the defences of the Rhine to the south-east or those of the Danube to the north-west. The Alamanni settled just outside the Roman fortifications of the Decumatian Fields and started to attack them during the reign of Caracalla. The emperor was able to contain their military activities by improving the military infrastructure in the Decumatian Fields, building many new roads, but this had a very significant financial cost for the Roman state. The sudden death of Caracalla had extremely negative consequences for the military situation of the Empire, both on the northern borders – where the Alamanni could reorganize themselves in view of further incursions – and in the Middle East. The leader of the Praetorians who had arranged the assassination of Caracalla, Macrinus, was proclaimed emperor in the Middle East but had to continue the war against the Parthians. The conflict came to an end only after the bloody Battle of Nisibis, which took place during 217 and resulted in the Romans being defeated after three days of harsh fighting. Macrinus was greatly weakened by this significant setback. Furthermore, he soon had to face a revolt by Elagabalus, who was a cousin of Caracalla. Elagabalus, thanks to the support of the Senate, dethroned Macrinus and rose to power. The reign of Elagabalus, however, was only brief, since in 222 he was assassinated and replaced by his young nephew Severus Alexander.

The Battle of Nisibis was the last clash fought between Parthians and Romans. Soon after it, the Parthian Empire crumbled and the new Persian dynasty of the Sassanids rose to power in present-day Iran. The Sassanids had ambitions of expelling the Romans from the Middle East, and Emperor Severus Alexander was soon forced to face them on the field of battle. The Roman Army campaigned against the Sassanids in 231 and 232, obtaining some successes but being unable to secure a decisive victory. Severus Alexander's war against the new Persian Empire proved to the Roman leaders that the Sassanids were much more powerful and dangerous than their Parthian predecessors. In 235, the emperor had to face a massive invasion of the Alamanni on the northern *limes*, but before he could face these enemies he was killed by his own soldiers, who no longer considered him to be a capable military leader. The assassination of Severus Alexander, as we have seen, marked the beginning of the terrible Crisis of the Third Century for the Roman Empire.

In 235, the leadership of the Roman Empire was taken by Maximinus Thrax, who commanded one of the legions fighting against the Alamanni when Severus Alexander was killed. The new ruler, being an experienced military commander, obtained a series of significant victories over the Alamanni in the Decumatian Fields. As soon as the situation on the Rhine had been stabilized, however, he had to move to the Danube in order to counter the incursions of the Sarmatians. During 238, a major internal revolt broke out in Roman Africa, where a local governor named Gordian was proclaimed emperor by his troops. Maximinus Thrax was able to rapidly crush the rebellion, but in doing so completely lost the support of the Senate. The senators revolted against his rule and proclaimed as emperor the grandson of Gordian, Gordian III. Maximinus Thrax then tried to invade the Italian peninsula from the north by moving from his strongholds in Pannonia, but while besieging the strategic city of Aquileia he was killed by his own soldiers. Gordian III ruled for less than two years and was unable to deal with the new threats that were emerging on the borders of the Empire. He had to fight a new war against the Sassanids, who were guided by the ambitious Shapur I. Initially, he obtained some significant victories, for example at the Battle of Resaena, but in 244 he was defeated and killed at the Battle of Misiche. Gordian III's successor was the commander of the Praetorians, Philip the Arab, who soon made peace with Shapur I but then had to face the Germanic tribe of the Carpi on the Danube. In 249, after hostilities with the Sassanids had resumed and the Roman defences on the Danube had been broken at various points, the legions of Pannonia acclaimed Decius – who had been one of Philip the Arab's main

Reconstruction of the costume worn by Maxentius at the decisive Battle of the Milvian Bridge. (*Photograph by Rachele Lori, copyright by Protectores Domini Nostri*)

Reconstruction of the costume worn by Maxentius at the Battle of the Milvian Bridge. The helmet shows a clear Hellenic influence, which was popular among the high-ranking officers. (*Photograph by Rachele Lori, copyright by Protectores Domini Nostri*)

Guardsman of the *Protectores* (left) and officer (right) from the years of the Tetrarchy. Both are wearing ridge helmets; the guardsman is heavily armoured with scale cuirass. (*Photograph by Rachele Lori, copyright by Protectores Domini Nostri*)

collaborators – as the new emperor. The troops still loyal to Philip were defeated in northern Italy, allowing Decius to assume power in Rome.

Decius was the first Roman monarch to face the Goths on the Balkan frontier, but could do little to stop their raids and incursions. In 251, a massive clash took place between the armies of the Romans and the Goths at the Battle of Abritus. Despite suffering severe losses, the Goths emerged victorious. Decius, after having fought with courage, was killed on the field of battle: he was the first Roman emperor to die during a battle fought against a foreign enemy. Decius' successor was Trebonianus Gallus, who soon made peace with the Goths in order to stabilize the Danubian frontier. During his two-year reign, the new emperor had to deal with two consecutive Sassanid invasions of his eastern provinces and several barbaric incursions across the Danube. The Roman troops, dissatisfied with the leadership of Trebonianus Gallus, rose up in revolt, the ensuing civil war ending only when the emperor was betrayed and assassinated by his own supporters. Following these events, the Roman Empire entered a period of complete political chaos. Aemilianus, a commander of the Roman troops stationed in the northern Balkans, ruled as emperor for just two months before being killed by his own soldiers, who favoured the ascendancy of another general – Valerian – to the throne. The latter ruled until 260 and spent most of his reign fighting against the Sassanids of Shapur I, who had occupied the Roman Middle East. Valerian could count on the support of his son, Gallienus, who was a very capable military commander. Despite this, however, he was soundly defeated by the Sassanids at the Battle of Edessa in 260. This was one of the worst military disasters ever suffered by the Romans, involving the destruction of a large part of the Roman army and the capture of Valerian. The emperor, after suffering humiliations at the hands of Shapur I, died – probably in 264 – as a captive of the Sassanids.

The defeat and capture of Valerian led to the beginning of the worst and most chaotic phase of the Crisis of the Third Century. Valerian's son and successor Gallienus, had to face the outbreak of a series of internal revolts. In the eastern provinces, which were no longer protected by Roman troops and thus were now exposed to Sassanid conquest, the Roman client king of Palmyra – Odaenathus – assumed power over all the lands that had previously been under imperial control. At the same time, in the north, Postumus, the commander of the troops stationed on the Rhine frontier, rebelled against Gallienus. Postumus soon assumed control over most of the Empire – Gaul, Britannia and Iberia – and then seceded his territories from the Roman state. He did not make any effort to depose Gallienus in Rome. Instead, he created an autonomous state known as the Gallic Empire, which had its own parallel institutions modelled on those of the Roman Empire's central government. The Gallic Empire had its own army, Praetorian Guard, coins, consuls and Senate. Both Postumus and Odaenathus proved to be very skilful military leaders. Postumus decisively defeated both the Alamanni and the Franks, restoring order in the Rhineland, while Odaenathus overcame Shapur I and freed the Roman Middle East (Syria and Mesopotamia) from Sassanid rule. In 262, Odaenathus even besieged the Sassanid capital of Ctesiphon before being proclaimed King of Kings of the East.

During 267, Odaenathus was assassinated, for reasons that remain unknown, and was succeeded by his infant son, Vaballathus, who reigned under the regency of his mother, Zenobia. Postumus, meanwhile, was killed during a rebellion of his own troops in 269. While these events took place in the provinces that had seceded from the Empire, Gallienus struggled to defend the Italian peninsula and the Balkans from the destructive incursions of the Germanic peoples. Thanks to his intelligent reform of Rome's forces, he was able to obtain some significant victories over the invaders. In 268, however, his cavalry commander, Aureolus, revolted against him and a new civil war broke out, which ended with the defeat of Aureolus but also with the assassination of Gallienus. The emperor was succeeded by Claudius Gothicus, an experienced military commander who obtained a series of great victories. In 268, he defeated the Goths at the Battle of Naissus and then the Alamanni at the Battle of Lake Benacus. The following year, Claudius Gothicus reconquered Iberia from the Gallic Empire, but he died from smallpox during the early months of 270, after having appointed his cavalry commander – Aurelian – as his successor.

Aurelian was without doubt one of the greatest Roman rulers, as in a reign lasting just five years he was able to reunify the Empire and temporarily stabilize its borders. First of all, he once again defeated the Alamanni and the Goths in order to secure his position in Italy and the Balkans. He then decided to evacuate the province of Dacia, which was too exposed to foreign invasions due to it being located north of the protection of the Danube. In 272, Aurelian turned against the newly formed Palmyrene Empire ruled by Queen Zenobia, which comprised much of the Middle East: south-eastern Anatolia, Syria, Palestine and even Egypt. Aurelian, recognizing that Rome could not survive without the massive amounts of grain produced in Egypt, organized a massive offensive against Palmyra. Moving across Anatolia, the Roman troops reconquered one province after the other, obtaining several spectacular victories. They then besieged the city of Palmyra, which was captured together with Zenobia. Palmyra was razed to the ground, while Zenobia was paraded in chains during Aurelian's triumph in Rome. In 274, the emperor turned his attention to the destruction of the Gallic Empire. He assembled a large army and invaded Gaul, hoping to force a major pitched battle. At the subsequent Battle of Chalons, which took place in northern France in early 274, Tetricus, the new leader of the Gallic Empire, was severely defeated and forced to surrender. Soon after these events, the Gallic Empire dissolved and its territories were re-annexed by the Roman Empire. The following year, while preparing a campaign against the Sassanid Empire, Aurelian was assassinated by one of his supporters. The unexpected death of Aurelian opened a new phase of political chaos, which saw the election of two emperors – the half-brothers Tacitus and Florianus – who reigned for just a few months. In June

Officer from the years of the Tetrarchy with a corselet of scale armour. (*Photograph by Rachele Lori, copyright by Protectores Domini Nostri and Legio Palatina Ioviani Seniores*)

Officers from the years of the Tetrarchy. The figure on the right is off-duty, as clear from his clothes and from his *pileus pannonicus* headgear. (*Photograph by Rachele Lori, copyright by Protectores Domini Nostri*)

Legionary from the mid-third century with lorica hamata. (*Photograph and copyright by Varianus/ Numerus Brittonum*)

276, Probus, who had been one of Aurelian's most important military commanders, became emperor, with the support of the Senate. Probus was a very capable military leader, achieving many victories over the Alamanni, Franks, Burgundians and Goths. Probus campaigned on the territories of his Germanic enemies and repaired the Roman defensive structures in the Decumatian Fields. In 282, however, he was assassinated as part of a military coup. His successor was Carus, who ruled for just a few months, during which he was able to obtain victories over both the Sarmatians and the Sassanids. Carus probably died of natural causes and was succeeded by his son, Carinus. The latter, however, soon had to face the ambitious Diocletian, who was proclaimed emperor by the Roman troops of the eastern provinces in 284. At the Battle of the Margus in the summer of 285, Diocletian defeated Carinus, who was assassinated shortly afterwards by one of his officers, leaving his rival as the supreme ruler of the Roman Empire.

As we have seen, Diocletian put an end to the Crisis of the Third Century by creating the new political structure known as the Tetrarchy. Being a veteran military commander of humble origins, he stabilized, together with his loyal co-emperor Maximian, the frontiers without conducting any major military campaigns. Diocletian ruled over the portion of the Empire in the East and thus faced the Sassanids, while Maximian reigned in the West and fought against the Germanic peoples. In 293, according to the new system of the Tetrarchy, both Diocletian and Maximian appointed their 'junior emperors' who were, respectively, Galerius and Constantius Clorus. The latter pair enjoyed mixed fortunes in war: while Galerius was defeated by the Sassanids, Constantius Clorus obtained some significant victories over the Picts in Britain. In 305, Diocletian and Maximian jointly abdicated and retired, thereby elevating Galerius and Constantius Clorus to the rank of senior emperors. They in turn soon appointed their own junior emperors: Maximinus Daza in the eastern territories for Galerius, and Severus in the western territories for Constantius Clorus. However, Constantius Clorus died in York while campaigning in 306, causing the first significant problems for the new system of the Tetrarchy, since rather than accepting the elevation of Severus from junior to senior emperor, the Roman legions in Britannia proclaimed Constantius Clorus' son, Constantine, as new senior emperor. Galerius tried to avoid the outbreak of a civil war between Severus and Constantine by assigning the position of senior emperor to the former and junior emperor to the latter.

The actions of Galerius temporarily avoided the outbreak of a civil war, but were interpreted – correctly – as a violation of the Tetrarchy's rules by Maximian's son, Maxentius, who declared himself emperor at Rome in 306, initiating what became known as the Civil Wars of the Tetrarchy. Severus moved from his capital

Legionary from the mid-third century with Niederbieber helmet. (*Photograph and copyright by Varianus/ Numerus Brittonum*)

Legionary from the mid-third century with *lorica hamata*. (*Photograph and copyright by Varianus/ Numerus Brittonum*)

Legionary from the mid-third century with Niederbieber helmet. (*Photograph and copyright by Varianus/ Numerus Brittonum*)

of Mediolanum against Rome, at the head of the army that had previously been commanded by Maximian. However, when Severus besieged Rome, his troops deserted and swore loyalty to Maxentius. The latter forced Severus to surrender and started to rule the Italian peninsula jointly with his father, Maximian, who returned to active political life. In 307, Galerius marched to Italy at the head of a large army, but his invasion was an abject failure, Maxentius being able to win the loyalty of his opponent's soldiers and to convince them to mutiny. Galerius, defeated, was forced to return to his eastern territories. Following these events, Maximian tried to establish an alliance with Constantine and arranged the marriage of the former junior emperor with his daughter, Fausta. In 308, a split developed between Maximian and his son, Maxentius, with Maximian trying unsuccessfully to eliminate his direct heir. At this point, the elderly Maximian fled north and joined Constantine, being in search of an ally who could defeat Maxentius. Shortly after these events, Licinius was appointed by Galerius as the new senior emperor of the West and was ordered to defeat the usurper Maxentius as soon as possible. In 310, Maximian rebelled against his protector, Constantine, and tried to cause a mutiny of the latter's troops, but he was soon defeated and captured by Constantine, who suspended his military operations against the Franks to move against him. Maximian died shortly afterwards, having been encouraged by Constantine to commit suicide. Galerius died in 311, destroying what little stability remained in the Tetrarchy. Maximinus Daza and Licinius soon started fighting against each other for dominance over the eastern part of the Roman Empire. Meanwhile, Maxentius mobilized his forces for a direct confrontation with Constantine. During the winter of 311/312, Constantine, a skilled diplomat as well as a brilliant military commander, forged a strong alliance with Licinius, to whom he arranged the marriage of his sister, Constantia. In response to these moves, Maximinus Daza allied himself with Maxentius and sent some of his troops to Italy.

Constantine swiftly prepared a pre-emptive attack on Maxentius, which materialized in the spring of 312 when he crossed the Alps at the head of 40,000 soldiers. Maxentius tried to slow down the advance of his enemy by sending cavalry forces to northern Italy, but they were routed by Constantine, who moved rapidly to occupy the important city of Mediolanum. Maxentius' military forces held very strong defensive positions around present-day Verona, but these were destroyed without great difficulty by Constantine. Verona was occupied by the invaders, together with other important centres of northern Italy such as Aquileia and Ravenna. Thanks to his brilliant military successes in northern Italy, Constantine could now enter central Italy and march on Rome. Maxentius prepared himself for a siege in his capital: he could still count on the loyalty of the Praetorian Guard and was well stocked with African grain. Rome was surrounded by the recently built and seemingly impregnable

Legionary from the mid-third century with Niederbieber helmet of the Frankfurt type. (*Photograph and copyright by Varianus/Numerus Brittonum*)

Legionaries from the mid-third century equipped with spear and oval shield. (*Photograph and copyright by Varianus/Numerus Brittonum*)

Aurelian Walls, meaning it would have been very difficult for Constantine to take the city. Maxentius ordered the destruction of all the bridges across the Tiber and built a single boat bridge across the river. While Constantine marched across central Italy, however, Maxentius was forced to change his strategy due to increasing malcontent among Rome's inhabitants. Recognizing that in the event of a siege the population would revolt against his rule and side with Constantine, he decided to assemble all his troops and move against his rival in order to fight a decisive battle. This battle

Legionary from the mid-third century wearing *lorica hamata*. (*Photograph and copyright by Varianus/Numerus Brittonum*)

took place not far from the new boat bridge – known as the Milvian Bridge – that Maxentius had recently built. On paper, he could count on an army that was much larger than that of his rival, but Constantine's troops consisted of battle-hardened veterans. In addition, Constantine boosted the morale of his men by giving them a new insignia with the 'Chi-Rho' Christian symbol shortly before the clash. According to legend, Constantine was visited by a dream the night before the battle and was advised to adopt such a symbol in order to achieve victory (in the dream, Constantine was told the famous expression *In Hoc Signo Vinces*, 'Win By This Symbol'). The Greek letters Chi (X) and Rho (P) were the first two letters of the Greek spelling of the word 'Christ', and together they were used as a very common Christian symbol, so Constantine's decision to have them on his new insignia had a very clear religious meaning. Indeed, before facing his fiercest enemy in a decisive engagement, he embraced the Christian faith. Constantine's conversion was perceived as an example to follow by many of his soldiers, marking the beginning of a new era in Roman history. If Constantine prevailed in the upcoming battle, the Christian faith would be recognized as a legitimate religion of the Roman Empire, after having been persecuted for centuries (especially under Diocletian, who was responsible for the murder of thousands of Christians). From an ideological point of view, the Battle of the Milvian Bridge thus became a clash between the new champion of the Christians (Constantine) and the last defender of paganism (Maxentius).

The Battle of the Milvian Bridge began with a charge by Constantine's cavalry, which rapidly destroyed the mounted forces of Maxentius. After this initial success, Constantine ordered a general advance of his infantry: his battle-hardened veterans, thanks to their superior training and discipline, crushed the enemy lines with a single charge and pushed many of Maxentius' soldiers into the Tiber, where they were killed or drowned. The Praetorians were the only soldiers of Maxentius who kept their position, but after being charged by the cavalry of Constantine even they broke ranks and fled to the river. Maxentius, accepting that the battle was lost, attempted to cross the Milvian Bridge in order to reach Rome, but he was pushed into the Tiber by the mass of his fleeing soldiers and drowned. Constantine's victory was complete. He was met with popular jubilation upon entering Rome, and was thus able to secure control over the whole of Italy. To win the loyalty of the African territories, who had sided with Maxentius, he sent the head of Maxentius to Carthage. Fearing punishment by the new master of Rome, the military commanders in Africa joined Constantine's cause.

While these events took place in the western territories of the Empire, in the East, Maximinus Daza and Licinius confronted each other. Maximinus Daza exerted control over the Asiatic territories, while Licinius secured his power over the Balkans. In 313, Maximinus Daza left Syria at the head of 70,000 soldiers and reached Anatolia, with the objective of landing in the Balkans. Harsh weather he encountered along the way, however, severely weakened his army and gave precious time to Licinius to strengthen his forces. In April 313, Maximinus Daza crossed the Bosporus and went to Byzantium, one of Licinius' strongholds. The city was taken after a brief siege, and the same happened a few days later at Heraclea. Licinius, having been able to assemble an army of 30,000 men, moved against his rival not far from Heraclea. The decisive clash of the campaign took place at Tzirallum, where – against all odds – Licinius achieved a clear victory. The defeated Maximinus Daza had no choice but to abandon his invasion of the Balkans and return to his Asiatic territories. He tried to halt Licinius' advance at the Cilician Gates, a strategic system of mountain passes that connected southern Anatolia with northern Syria. Licinius, however, broke through the hastily prepared defences of his opponent and forced him to retreat to Tarsus. Maximinus Daza then died (probably of natural causes, but some contemporaries suggested that he had been poisoned). This unexpected event led to the end of the hostilities and left Licinius as the only ruler of the Roman Empire's eastern half. By the closing months of 313, a military confrontation

Legionary from the mid-third century wearing Niederbieber helmet. (*Photograph and copyright by Varianus/Numerus Brittonum*)

Legionary from the mid-third century with mail cuirass. (*Photograph and copyright by Paul Elliott*)

Legionary from the mid-third century equipped with spear and oval shield. (*Photograph and copyright by Paul Elliott*)

between Constantine and Licinius seemed inevitable, both leaders being determined to end the political system of the Tetrarchy and reunify the Empire under their own control. There was no immediate clash, though, with both rulers reorganizing their forces and consolidating their power. In 313, before Licinius could assume complete control of the East, the pair had met in Milan (Mediolanum) to confirm the alliance existing between themselves and had agreed to enact the Edict of Milan. This, for the first time in Roman history, officially granted tolerance to Christianity and to all the other religions of the Empire. It repudiated any form of religious coercion and greatly favoured the spread of the Christian faith.

Eventually, in 316, hostilities broke out between Constantine and Licinius. The war began with a massive offensive by Constantine in the Balkans, where Licinius, being in numerical inferiority, had no choice but to fall back in good order. Licinius inflicted significant losses on his opponent and prevented the disintegration of his own army. Constantine's offensive had started with a significant victory in Pannonia, at the Battle of Cibalae, but Licinius' forces had not been destroyed there. After months of campaigning in the Balkans, another major clash took place at Mardia in Thrace. This proved to be extremely bloody for both sides and ended in another indecisive victory for Constantine, Licinius managing to escape with a large number of his troops. At this point in the campaign, both sides were exhausted and decided to suspend hostilities with the signing of a peace treaty. According to its terms, which were agreed in March 317, Licinius ceded to Constantine all his Balkan territories except for Thrace. The following years were spent by both emperors in preparation for a further war, one which would prove decisive for the destiny of the Roman world.

In 320, Licinius reneged on the religious freedom promised by the Edict of Milan and resumed the persecution of Christians in his territories. These acts were perceived as a challenge by Constantine, who was determined to use Christianity as an innovative social element that could aid the survival of the Roman Empire. In 324, hostilities between Constantine and Licinius resumed. Constantine moved against his rival's defensive positions located on the frontier of Thrace, and the first significant engagement of the new conflict was fought at Adrianople. This again ended in victory for Constantine. The Battle of Adrianople was a gigantic clash, pitting 130,000 soldiers of Constantine against 160,000 men of Licinius. Both armies comprised significant numbers of Germanic warriors, since Constantine had allied himself with the Franks and Licinius had secured strong support from the Goths. Following the clash, Constantine was able to expel Licinius from his last territorial possession in Europe and conquer Thrace. Before crossing the sea to invade Asia, Constantine had to send his fleet against that of Licinius. In the ensuing Battle of the Hellespont, one of the largest naval clashes of Antiquity, Licinius' fleet

Legionary from the mid-third century with Heddernheim Type E helmet. (*Photograph and copyright by Paul Elliott*)

Legionary from the mid-third century wearing metal greaves. (*Photograph and copyright by Paul Elliott*)

Legionary from the mid-third century wearing *lorica hamata*. (*Photograph and copyright by Paul Elliott*)

– despite its clear numerical superiority – was completely destroyed. Following this success, Constantine could take the strategic city of Byzantium and land in Anatolia. He then marched towards Chalcedon, pressing for the pitched battle that would finally determine the outcome of the war. The decisive Battle of Chrysopolis was fought on 18 September 324. It began with a massive frontal attack launched by Constantine, which rapidly routed most of Licinius' troops. Constantine's soldiers killed around 30,000 of their opponents. Licinius fled to the city of Nicomedia, where he finally surrendered to Constantine on the promise that his life would be spared. Constantine accepted such terms and permitted to Licinius to live as a private citizen. Nevertheless, in 325, Constantine accused his old rival of plotting against his rule and had him executed. The death of Licinius left Constantine – who was later nicknamed Constantine the Great – as the sole ruler of the reunified Roman Empire. The Civil Wars of the Tetrarchy had ended and the new historical phase known as the Late Empire had officially began. Constantine's reign saw the consolidation of the Empire's borders, with the temporary reconquest of the Decumatian Fields from the Alamanni and of part of Dacia from the Sarmatians. Constantine repulsed the Goths north of the Danube and even planned to conduct a major invasion of the Sassanid Empire. Before such a scheme could materialize, however, he died of natural causes in 337. Following his death, the Roman Empire was partitioned between his three sons – Constantine II, Constantius II and Constans – until 350, when the last survivor, Constantius, fought another Roman civil war against the usurper Magnentius, who had ordered the assassination of Constans. Constantius eventually emerged victorious and reigned until his death in 361.

Chapter 4

The Military Reforms of Constantine

The long Crisis of the Third Century ended only with the ascension to power of Diocletian in AD 284. After decades of internal unrest, civil wars, secessions and foreign invasions, the general situation of the Empire was very dramatic. The Roman civilization had survived and the Empire had retained territorial unity, but all the institutions of the state were experiencing severe difficulties. Diocletian soon understood that the political and military structures of the Empire had to be completely reformed in order to survive: the political organization based on centralism and the military organization based on the legions were of no use against the new internal and external threats that were menacing the stability of the Empire. First of all, Diocletian decided to find an alternative to the excessive centralism of the Roman administrative apparatus that had been among the main causes behind the ineffective military response to the foreign invasions and to the secessions of Gaul and Palmyra. He then gradually developed the innovative system known as the Tetrarchy, which was created with the main aim of avoiding succession wars when an emperor died. Since the dynastic system of succession had proved to be totally inadequate for the new times, Diocletian decided to start determining imperial successions according to the 'adoption principle'. Each ruling emperor was now to choose his own heir soon after assuming power, in order to avoid internecine conflicts: future emperors had to be chosen according to their personal valour and military virtues, for the benefit of the Roman state. In addition to this, the whole Empire was divided into four parts, ruled by two senior emperors and two junior emperors. Each senior emperor, known as Augustus, controlled one half of the Empire, and they elected their own successor, who were known as Caesar (junior emperor). Each half of the Empire was divided in two parts: one ruled by the Augustus and the other by the Caesar. The senior emperor had a certain formal control over the junior one, but in general it can be said that the Empire was effectively divided into four smaller sub-empires with autonomous administrative structures. The Tetrarchy of Diocletian was a very intelligent response to several political problems, first of all that of successions. When an Augustus died or retired, the Caesar whom he had chosen took his place and nominated a new junior emperor. The idea was that there would be no more space for civil wars. Similarly, the causes of secessions had been practically cancelled:

each of the four parts of the Empire had its own supreme leader and administrative apparatus, thus augmenting the efficiency of defence in case of foreign invasion. The old administrative division based on provinces was abolished, being substituted with a new one that saw the organization of the territory into twelve dioceses: the four parts of the Empire ruled by the tetrarchs were known as praetorian prefectures, and each of which was to include three dioceses. All these changes greatly simplified the internal organization of the Empire, especially from a military point of view, with flexibility and adaptability being the key elements for success and the only way to face multiple threats. Each of the dioceses was ruled by a governor known as a *vicarius*, while each praetorian prefecture was governed by an Augustus or Caesar, with the decisive support of a chief-of-staff known as a *praefectus praetorio*. From a military point of view, the reforms of Diocletian were revolutionary and highly effective. First of all, he decided to separate in a definitive way the military apparatus from the administrative setup. He did this in all sectors of the state, from the lowest to the top of the imperial command structure. For centuries, the senatorial class had controlled both the military and civil administrations, but the specific needs of the new times had gradually shown all the deficiencies of this system. Both the army and the civil administration now required high levels of specialization, which forced Diocletian to create separate positions and responsibilities. The military was in the hands of the *equites*, who had already been in command of all the major units since the time of Gallienus, while the civil administration remained under the control of the senatorial class, which now had more specific economic and social functions.

As we have seen in the previous chapter, Constantine became supreme ruler of the Western Empire in 312 after defeating Maxentius at the Battle of the Milvian Bridge. Some years later, he also defeated the legitimate eastern emperor, Licinius, thereby reuniting all the territories of the Roman Empire under the power of a single ruler. In many respects, Constantine can be considered the last great emperor of Rome, largely because his own personal conversion led to the adoption of Christianity as the official religion of Rome. It is the opinion of several historians, however, that his most important acts were his military reforms. In general terms, it can be said that Constantine was the first emperor since Augustus to completely change the structure of Rome's military forces. His first act in this sense, after defeating Maxentius, was to disband the Praetorian Guard: this was an epochal decision, one that nobody before Constantine had seriously attempted to do. For almost three centuries, the Praetorians had determined the destiny of Rome, by either killing emperors or electing new ones: now this incredibly powerful military institution was ended, as a result of the Praetorian Guard's choice to side with Maxentius against Constantine. The capital of the Empire was transferred by Constantine from Rome to the new

Legionary from the mid-third century with Niederbieber helmet of the Frankfurt type. (*Photograph and copyright by Varianus/Numerus Brittonum*)

Legionary from the mid-third century with short-sleeved corselet of mail armour. (*Photograph and copyright by Varianus/Numerus Brittonum*)

city of Constantinople that he had just founded, so the existence of a strong imperial guard based in Rome no longer made sense.

The corrupt Praetorians were replaced by new units of mounted guardsmen known as *Scholae Palatinae*, which guaranteed a highly mobile escort service to the Emperor even when he was on campaign (differently from the Praetorian Guard, whose cavalry was limited to the *Equites singulares Augusti*). At the time of Diocletian, the term *schola* was commonly used to indicate an organized component of the imperial retinue (civil or military). Each of the various components had its own specific chamber (*schola*) in the imperial palace. Later, the term started to be used to indicate the mounted regiments forming the imperial guard. Each *Schola Palatina* was an elite cavalry *vexillatio*, with the same internal organization as regular mounted units. As we will see later, under Constantine, the standard *vexillatio* had the same internal structure as the former *ala quingenaria* (sixteen *turmae* of thirty men). Each *Schola* was commanded by a *Tribunus Militum*, but unlike the Praetorian Guard, the *Scholae* had no overall military commander because the Emperor retained direct control over them. Initially, there were five *Scholae*, but when the Empire was again divided into two parts after the death of Constantine, their number

Legionary from the mid-third century equipped with spear and shield. Note the decorations painted on the back of the latter. (*Photograph and copyright by Varianus/Numerus Brittonum*)

was increased to ten (five for each half of the Empire). The original units started to be defined as *seniores*, while the newly formed ones became known as *iuniores*. These terms were largely also used after the definitive division of the Empire in 395, when many of the most important military units were duplicated: the original ones, generally remaining in the Western Empire, were called *seniores*, and the new ones were known as *iuniores*. By the time of the *Notitia Dignitatum*, the structure of the *Scholae Palatinae* had changed again, with eight units in the Western Empire and seven in the Eastern Empire. Since the *Scholae Palatinae* had the difficult function of protecting the emperor, their men were generally chosen among Germanic recruits: Franks and Alamanni in the Western Empire, Goths in the Eastern Empire. After the crushing Roman defeat by Gothic rebels at the Battle of Adrianople in 378, however, the Goth guardsmen of the eastern emperors were gradually substituted by recruits from Armenia and Isauria. The best elements of the *Scholae Palatinae*, comprising just forty men, were detached from their units in order to form the small inner guard of the emperor. Chosen according to their personal loyalty, these were known as *Candidati* because of their bright white tunics. They were commanded by an officer known as a *primicerius* and followed the emperor during every single moment of his day, including when on campaign. Due to their guard status, the members of the *Scholae Palatinae* had a series of privileges, like the former Praetorians: they received higher pay than normal soldiers and enjoyed special treatment including receiving extra rations of food and being exempt from paying some taxes). During the age of Constantine, the *Scholae Palatinae* were the elite of the Roman Army, but this situation gradually changed over time. The ease of palace life and lack of actual campaigning reduced the combat abilities of these guardsmen, transforming them into parade-ground soldiers. It even became possible to buy an appointment into the ranks of the *Scholae Palatinae*, a situation that would have been considered totally unacceptable at the time of Constantine.

After the abolition of the Praetorian Guard, Constantine organized another corps of imperial guardsmen that had different functions from the *Scholae Palatinae*. These were known as *Protectores* and served both as bodyguards and staff officers. They were all junior officers who were chosen by the emperor because of their superior personal abilities. They included both foot and mounted guardsmen, the latter also being known as *Domestici*, in order to distinguish them from the foot bodyguards, who were simply called *Protectores*. The commanding officer of the foot guards was known as the *comes domesticorum peditum*, while the commander of the mounted guards was the *comes domesticorum equitum*. All the *Protectores* were under the direct command of a senior officer known as the *comes domesticorum*. When the Empire was definitively divided in two, the number of *Protectores* were doubled, as were many

Legionary from the mid-third century, wearing metal greaves and *manica*. The latter consisted of segmented armour and was worn only on the right arm. (*Photograph and copyright by Varianus/ Numerus Brittonum*)

Vexillifer from the mid-third century, equipped with mail armour. (*Photograph and copyright by Varianus/ Numerus Brittonum*)

Vexillifer from the mid-third century, with oval shield. (*Photograph and copyright by Varianus/Numerus Brittonum*)

other units. Like the members of the *Scholae*, the *Protectores* were mostly Germanic in origin and generally came from the equestrian class. Because of their extreme loyalty, they were frequently used for special missions by the emperor, who sent them across the Empire as his own personal agents. When serving on such special missions, the *Protectores* were commonly defined as *deputati*. The corps of *Protectores* can be seen as a sort of general staff or military academy, service in the unit being used to create the most brilliant junior officers. After a few years of service in the *Protectores*, a member of the unit was generally granted a commission by the emperor and placed in command of a military unit. Originally, the title of 'protector' was introduced during the reign of Gallienus, when it was given to individual officers as a mark of their devotion towards the emperor (mostly *tribuni* or centurions). After being awarded with this honorific title, officers were likely to receive accelerated promotion in their service. In general terms, the original *Protectores* can be described as a military order attached to the general staff of the emperor. After the reign of Gallienus, the *Protectores* were gradually transformed into a proper military unit, in a process that was fully completed under Constantine.

After reforming the guard units, Constantine initiated a complete transformation of the whole Roman Army. The basic idea behind this general reform was to create a new fundamental distinction between military units, with the field armies separated from the border troops. The field armies were garrisoned in the major cities of the Empire, as highly mobile strategic reserves, while the border troops were stationed along the *limes*, serving as the first line of defence against foreign attacks. In practice, the reform was based on the new principle of defence in depth. In case of simultaneous attacks on various points of the border, the frontier troops were to stop the enemies for just a few hours or days, while the field armies would be rapidly assembled and sent against the invaders. This new system had a great advantage, as the presence of the field armies in the major cities of the Empire was very important for the internal security of the various territories because these elite troops could be swiftly used to crush rebellions or secessions. By the time of Constantine, internal stability was as important as defence against foreign invasions. As a result of these new principles, all the units of the Roman Army were divided into two large categories: the *comitatenses* (field troops) and the *limitanei* (border troops). The *comitatenses* were essentially the direct heirs of the former legions, while the *limitanei* had some characteristics in common with the former auxiliaries. The *comitatenses* were elite troops with excellent equipment and training, mostly legions or cavalry *vexillationes*, whereas the *limitanei* were organized into smaller units with a strong local character, usually showing a high degree of foreign influence (many units were equipped more or less like their direct enemies, similarly to the former *auxilia*). The *limitanei* were obviously much

Vexillifer from the mid-third century, with small round shield. (*Photograph and copyright by Varianus/ Numerus Brittonum*)

Legionary from the mid-third century wearing a piece of scale armour on the shoulders over the *lorica hamata*. (*Photograph and copyright by Varianus/Numerus Brittonum*)

Legionary from the mid-third century with corselet of scale armour. (*Photograph and copyright by Varianus/Numerus Brittonum*)

more static than the *comitatenses* and were strongly linked to the defensive positions that they had to protect (they were essentially a mix between border police and garrison militia).

The *limitanei* were designed to deal with minor enemy incursions or skirmishing activities. Indeed, most of the Roman border regions were characterized by an endemic state of low-intensity conflict, which the *limitanei* were prepared to face on a daily basis. In most cases, the *limitanei* had lighter equipment than the *comitatenses* and enjoyed fewer privileges (such as tax exemptions) than the field troops. At the end of their service, however, they received some land located near the border that they had defended for most of their lives. Analysing this important element, it is possible to suppose that Constantine wanted to create a whole new class of military settlers: the *limitanei* were defenders of the frontiers with strong personal links to the border lands of the Empire. This view is confirmed by the fact that by the time of Constantine's ascendancy to the throne, the profession of soldier had become mostly hereditary in the Roman Empire, so members of the *limitanei* would have passed their land properties to their sons and protection of the borders would be guaranteed by these military families. According to their particular functions, the limitanei could assume some special denominations: if they were stationed on the Rhine or Danubian frontiers, they were commonly known as *riparienses* (because their tract of frontier was marked by rivers); if they were garrisoned in large border forts, they were known as *castellani*; and if they were placed in smaller fortlets or watchtowers, they were known as *burgarii*. On special occasions, for example during a military emergency or for a large campaign, some units of *limitanei* could be attached to the *comitatenses*. In this case they were called *pseudocomitatenses*. Although their period of service with the field armies was generally quite limited, on some occasions the best units of *pseudocomitatenses* could remain attached to the field troops forever.

The comitatenses were grouped into several field armies, which were deployed across the regions of the Empire. There was, however, a special category of field army: the *comitatus praesentalis*, the emperor's own field army, which was under his direct control. Each of the other field armies was commanded by a high-ranking officer known as a *magister militum*, who was assisted by a *magister peditum* (commander of the infantry) and *magister equitum* (commander of the cavalry). Over time, the *comitatus praesentalis* started to be commanded by a *magister militum*, like all the other field armies. Smaller field armies, instead, started to be commanded by a senior officer known as a *comes*, and all the *limitanei* of a certain border region were under command of a *dux*. Both the *comes* and the *dux* were subordinate to the *magister militum*. The most senior units of the *comitatenses* field armies were known as *palatini*, which could be legions, cavalry vexillations or *auxilia palatina*, the latter being elite light infantry units created by Constantine by converting the semi-regular *numeri* of Germanic warriors into regular corps. The *auxilia palatina* were generally named after the barbarian tribe from which they had been recruited, but some could be named after the emperor who had established them. The Roman forces reformed by Constantine gradually started to include an increasing number of irregular military units composed of Germanic warriors. Generally speaking, the foreign irregulars serving within the Roman Army could belong to three different categories: the *buccellarii* (mercenaries), *laeti* (settler-soldiers) and the *foederati* (allies). After the military defeat at Adrianople against the Goths in 378, many leading Roman military commanders started to expand the forces under their orders by forming large contingents of Germanic mercenaries. These soldiers were recruited singularly or in groups and were assembled into units that were under the direct orders of their employers. Almost every *Magister Militum* started to have several units of mercenaries among his forces. These Germanic troops, paid directly by their employer, generally proved to be very loyal towards the various Roman officers but had no ties with the emperor or with the central administration of the state. In practical terms, it can be said that several private armies now flourished within the Roman Empire. Over time, the use of *buccellarii* became even more widespread, with many Roman nobles – especially in the provinces – starting to assemble security forces made up of Germanic mercenaries. Because the central state was in great difficulty, many rich landowners preferred to use foreign professional soldiers to protect their own private properties. The term *buccellarii* derives from the word *buccellatum*, a type of dry biscuit that was the ordinary ration of the Germanic mercenaries. The private bodyguards could occasionally be converted into regular military units, such as the *Placidi Valentiniaci felices*, who were originally recruited as *buccellarii* by Empress Galla Placidia and were later converted into an *auxilia palatina*. The units of *buccellarii* had no proper military organization, but during the later years of the Empire they became better trained and equipped than many regular units. If at the service of a rich general or noble, they were more motivated than regular soldiers.

The term *laeti* defines communities of barbarians that were permitted to settle on imperial territory in exchange for military service in the Roman Army. Initially, this recruiting system was used with defeated tribes who were absorbed into the Empire. The first *laeti* communities appeared during the reign of Marcus Aurelius, who granted some lands to the Germanic tribes defeated during the Marcomannic Wars. He did so because the Antonine Plague had depopulated large areas of the Empire, greatly reducing the traditional recruiting potential of Rome. This first experiment of Marcus Aurelius, however, was not a great success: the defeated Germanic warriors soon rose up in revolt and even seized, for a brief period, the

Legionary from the mid-third century with Weisenau helmet. (*Photograph and copyright by Varianus/Numerus Brittonum*)

important city of Ravenna. Each community of barbarians that wished to enter imperial territory had to conclude an agreement with the Roman authorities, which prescribed a series of obligations but sometimes also privileges. The *Notitia Dignitatum* lists several groups of *laeti*, who could also be known as *gentiles* (the two terms had no particular distinctions). The agreement granting land to a community of barbarians might specify a one-off contribution of recruits or a fixed number of recruits required each year. The lands given to Germanic settlers were known as *terrae laeticae* and were not part of the normal provincial administration, since they were under the direct control of a specific military officer (known as the *Praefectus Laetorum* or *Praefectus Gentilium*). There is still considerable dispute among scholars about the military structure of the *laeti*: it is not clear if these barbarian recruits formed their own distinct military formations or if they were simply absorbed into the existing units. The second hypothesis, however, seems to be the most plausible: the *Praefecti Laetorum* were just administrative officers controlling the communities of foreign settlers, but had no units of barbarians under their command. The *Notitia Dignitatum* contains two lists of *Praefecti Laetorum*, with one for Gaul and another for Italy.

Legionary from the mid-third century with Heddernheim Type E helmet. (*Photograph and copyright by Paul Elliott*)

Legionary from the mid-third century with scale armour and oval shield. (*Photograph and copyright by Paul Elliott*)

Foederati, meanwhile, were all those tribes that were allied to Rome due to the terms of a treaty (known as a *foedus*). Generally speaking, tribes of *foederati* were provided with benefits by Rome in exchange for military assistance. Their loyalty towards the Empire, however, did not run deep. Indeed, most of the early *foederati*, for example the Franks or the Goths, later transformed themselves into fully-independent 'barbarian' nations located within Roman territory. On some rare occasions, the military collaboration between Rome and the Germanic allies had positive results, but in most cases the Germanic tribes only accepted a *foedus* to gain entry into the Roman Empire and then created their own independent realm. The contingents of *foederati* were commanded by their own leaders and retained their peculiar military traditions.

During the Early Empire, both the legionaries and the auxiliaries were all *voluntarii* (volunteers). Although compulsory military conscription did exist in the Roman world, it was used only very rarely. The practice of conscription, which was known as *dilectus*, was employed only in case of military emergencies or during preparations for major campaigns for which large numbers of additional troops were required. This situation changed completely after the Crisis of the Third Century, and by the time of Constantine, the Roman Army relied mainly on compulsion for the recruitment of citizens. As we have seen, the reforms of Diocletian made military professions hereditary, with the sons of serving soldiers or veterans required by law to enlist in order to continue the service of their fathers. In addition to this, all citizens of the Empire were subject to a regular annual levy that was based on a principle known as indictio (a sort of land tax assessment). Each landowner living on the territory of the Empire was required to provide a certain number of recruits to the Roman Army, this number being defined by the amount of land tax that was due on each landowner's estates. Obviously, this second channel of recruitment was strongly contested by the various rich landowners, who preferred to keep their best men on their estates (for work or for their private defence). In order to replace their best men, landowners often sent the less fit or reliable men at their disposal; as an alternative, they could cheat the authorities by sending sons of soldiers (who were already obliged to serve) or vagabonds (known as *vagi*). As is clear from this general description, the principle of *indictio* could only be used in the countryside: it mostly involved peasant recruits and affected the local power of landowners. In the urban centres, this land-based conscription could not be applied. As a result, the cities made little contribution to the defence of the Empire. Landowners were frequently able to convert their conscription requirements into a cash levy, with a fixed rate per recruit due. This was known as the *aureum tironicum*, from the word *tirones*, which meant 'recruits'. Generally speaking, the various regions of the Empire contributed in different ways to the formation of

Legionary from the mid-third century with Niederbieber helmet and scale armour. (*Photograph and copyright by Varianus/Numerus Brittonum*)

Legionary from the mid-third century equipped with a corselet of scale armour, having two metal plates on the chest. (*Photograph and copyright by Varianus/Numerus Brittonum*)

Legionary from the mid-third century with Weisenau helmet. (*Photograph and copyright by Varianus/Numerus Brittonum*)

Rome's military forces. The traditional recruiting areas of Gaul and the Balkans, for example, continued to provide the bulk of the Roman Army's recruits.

During the Middle Empire, the traditional height and age requirements for Rome's armed forces were progressively abandoned, with the result that the general quality of soldiers saw a dramatic decline. After being accepted into the Roman Army, each new recruit was marked on one of his arms with a specific brand, which was done in order to discourage desertion, which became an enormous problem during the last decades of the Western Empire. Each new recruit was issued with an identification disc that was worn around the neck, and was also given a certificate of enlistment that was known as a *probatoria*. The best recruits were generally assigned to the *comitatenses*, while the men with an inferior level of fitness were allocated to the static units of *limitanei*. The standard period of service was twenty years in the *comitatenses* and twenty-four years in the *limitanei*, the changes from the time of the legionaries and *auxilia* having not been particularly significant. All the difficulties related to recruiting had obvious consequences on the strength of individual corps, which became notably smaller than previously, especially the legions. Furthermore, there was always a massive disparity between official and actual strengths, with units in the field being generally much smaller than their theoretical establishment.

With the great military reforms of Constantine, the legion remained the basic unit of the Roman infantry, although its numerical strength and internal composition were greatly changed. Until the Crisis of the Third Century, the legions had acted more or less as small independent armies, having their own cavalry and artillery, but they lost their peculiar character after the organizational reforms of Diocletian and Constantine and their strength was much reduced. Most of the legions created or reorganized by Diocletian and Constantine derived from former *vexillationes* (detachments of legions), and the average number of their soldiers was between 1,000 and 1,200. Furthermore, they no longer had their own cavalry or artillery components. The new legions of the Roman Army were commanded by a *Tribunus* from the equestrian class, who usually had passed through the ranks of the *Protectores*. The *Tribunus* was assisted by a *Primicerius*, his second in command, this term being a new version of the earlier *Primus Pilus*. The *Primicerius* was an ex-ranker with considerable length of service, who had entered the officer corps because of his combat experience as a veteran soldier but who had not been part of the *Protectores*. The previous internal organization on cohorts was completely abandoned, with each legion now structured on six *ordines*, each of which included two centuries of legionaries. As a result, each *ordo* numbered 200 men and the whole legion comprised some 1,200 soldiers. Each *ordo* was commanded by an officer known as a *Ducenarius*, who could also be known as a Senator if he had already been chosen for service in the *Protectores* and was thus waiting for a vacancy in the ranks of the imperial staff college. Despite its title, the rank of Senator had nothing to do with the senatorial class. The first *centuria* of each *ordo* was commanded by the *Ducenarius*, while the second *centuria* was under the orders of a *Centenarius* (the equivalent of the former centurion). Both the *Ducenarius* and the *Centenarius* directly appointed their adjutant and second in command, who was called a *Biarchus* and performed more or less the same command functions as the former *Optio*, such as remaining at the back of the century during combat in order to keep the unit's close formation and discourage deserters from abandoning the field. As before, each century was structured on ten *contubernia*, each of which was now commanded by an NCO known as a *semissalis*, the equivalent of the former *decanus*. The majority of the legions were part of the *comitatenses* troops, but some of them were part of the limitanei.

As previously stated, the *auxilia palatina* were elite units of light infantry, generally performing special duties and having peculiar equipment. Their organization was quite similar to that of the heavy infantry legions, but each unit numbered just 500–600 men. Each *auxilia palatina* included just three *ordines*, with two centuries each. The chain of command and the internal organization of each *ordo* were exactly the same as in the legions. The majority of the *auxilia palatina* comprised light infantry

Legionary from the mid-third century wearing a piece of scale armour on the shoulders over the *lorica squamata*. (*Photograph and copyright by Varianus/Numerus Brittonum*)

Legionary from the mid-third century with scale corselet having two metal plates on the chest. (*Photograph and copyright by Varianus/Numerus Brittonum*)

Legionary from the mid-third century with Weisenau helmet. (*Photograph and copyright by Varianus/Numerus Brittonum*)

armed with small shields and javelins, but some units were entirely formed by foot archers.

Among the legions of Constantine's Roman Army, two had a special status: the *Ioviani* and the *Herculiani*, which were later split between *seniores* and *iuniores* with the definitive division of the Empire. After ascending to power, Diocletian and his co-emperor, Maximian, were respectively equated to the Roman gods Jove and Hercules; in order to secure their position and counter the power of the Praetorian Guard stationed in Rome, they promoted two faithful legions from the region of Illyricum (the *Legio V Iovia* and *Legio VI Herculia*) to be their personal guards. As a result of their promotion, the two legions were renamed *Ioviani* and *Herculiani* (from the honorific titles of Diocletian and Maximian). With Constantine's reforms, the two legions lost their guard status but continued to be considered as elite units.

Most of the *limitanei* units were structured as independent cohorts. The cohort of Constantine's Roman Army included six centuries of eighty men each, for a total of 480 soldiers. Each cohort was commanded by a *Tribunus*, who was assisted by a *Primicerius*, as in the legions. Each century was under the orders of an *Ordinarius*, who was the equivalent of the *Ducenarius* or *Centenarius* from the legions (but had fewer privileges than

Legionary from the mid-third century equipped with *manica* of segmented armour and metal greaves. (*Photograph and copyright by Varianus/ Numerus Brittonum*)

Legionary from the mid-third century equipped with extra-long scale corselet. (*Photograph and copyright by Varianus/Numerus Brittonum*)

them). Each *Ordinarius* was assisted by a *Biarchus* and commanded ten *contubernia*. The *numeri* continued to be units mostly formed by barbarians and retained their characteristic local nature as part of the *limitanei*; they were organized on eight centuries with eighty men each (for a total of 640 soldiers). Each *numerus* was commanded by a *Tribunus*, who had a *Primicerius* as his second in command. The eight centuries of each *numerus* were assembled into four *ordines*, each *ordo* being commanded by a *Ducenarius*, but differently from what happened in a legion, each century of the *numerus* was commanded by its own *Centenarius* (the *Ducenarius* did not command the first century of each *ordo*). To sum up, it can be said that the infantry of the *comitatenses* was organized into legions and *auxilia palatina*, whereas the infantry of the *limitanei* was formed into cohorts and *numeri*.

The basic cavalry unit of the *comitatenses* was the *vexillatio*, which initially had the same internal structure as the light infantry *auxilia palatina*. It included some 500–600 cavalrymen, structured on three *ordines* with two centuries each. The chain of command and the internal organization of each *ordo* were exactly the same as in the legions. With the definitive subdivision of the Roman Empire,

Legionary from the mid-third century wearing an additional piece of scale armour for protection of the shoulders. (*Photograph and copyright by Varianus/Numerus Brittonum*)

however, most of the cavalry *vexillationes* were divided in two like most of the other units. After having been halved, the *vexillationes* did not recruit back up to their original strength and kept their reduced numerical force, with just 300 horsemen each. Like the infantry units, the *vexillationes* of the Western Empire were generally known as *seniores*, while those of the Eastern Empire were called *iuniores*. Since each cavalry vexillatio now numbered only 300 men, the internal organization had to be changed: each unit now included ten *turmae* of thirty men. Each *vexillatio* was commanded by a *Tribunus*. The cavalry of the *limitanei* comprised two kinds of units: the *alae* and the *cunei*. While the border infantry retained the cohort, the border cavalry preserved the *ala*, which comprised 600 horsemen and was structured on twenty *turmae* of thirty men each. Each *ala* was led by a *Tribunus*. As already stated, the *cuneus* was originally created as a semi-regular unit of barbarians with no proper internal organization, but it gradually acquired a more stable structure. The *cunei* continued to be units mostly formed by barbarians and retained their characteristic local nature as part of the *limitanei*, but they started to be organized on ten *turmae* of thirty men, like the *vexillationes* of the *comitatenses* cavalry. Each *cuneus* was commanded by a *Tribunus*.

It is interesting to note that during this period of great change, the Roman Army started to have some independent

Legionary from the mid-third century wearing an additional piece of scale armour for protection of the shoulders. Note the decorative metal plates applied on the chest. (*Photograph and copyright by Varianus/Numerus Brittonum*)

Legionary from the mid-third century wearing a full set of scale armour. The latter could be silvered or gilded, as shown here. (*Photograph and copyright by Varianus/Numerus Brittonum*)

Legionary from the mid-third century wearing a scale corselet. Scale armour became immensely popular during the Crisis of the Third Century. (*Photograph and copyright by Varianus/Numerus Brittonum*)

artillery units, probably the first in the military history of the world. Until the Crisis of the Third Century, each legion deployed its own artillery, with each cohort having a stone-throwing *ballista* and each century having a bolt-shooter *scorpio*. A single Roman legion could deploy an impressive artillery detachment of ten heavy *ballistae* and fifty-nine light *scorpiones*. Around 250, the *ballistae* of each cohort started to be replaced by a heavier model of stone-thrower known as an *onager*, a catapult specifically designed for siege operations. The artillery of each legion was manned by very few specialists, since most of its crews were provided by the ordinary rankers. After the reforms of Constantine, however, this situation was completely changed: the new smaller legions did not have their own artillery detachments, as all the legionary artillerymen from the various units were assembled together to form new and independent artillery legions. These specialized units were known as *ballistarii*, and it seems that initially there was just one such specialist legion, which was part of the *comitatenses*. Later, with the division of the Empire, such units were divided between *seniores* and *iuniores*, as with many other corps. By the time of the *Notitia Dignitatum*, however, the Roman field armies included six artillery legions: the *Ballistarii seniores*, *Ballistarii iuniores*, *Ballistarii Dafnenses*, *Ballistarii Theodosiaci* and *Ballistarii Theodosiani iuniores* in the Eastern Empire, and the *Ballistarii* in the Western Empire. Only three of these units were *comitatenses*: the *Ballistarii Dafnenses* and the two original artillery legions (the *Ballistarii seniores* and *Ballistarii iuniores*). The remaining three artillery units were *pseudocomitatenses* and had been promoted in order to serve with the field armies. It is highly probable that these units were formed by gathering together garrison artillerymen who manned the defensive fortifications located on the *limes*; the *Notitia Dignitatum*, in fact, also lists an artillery unit of *limitanei* known as the *Milites ballistarii*.

Chapter 5

Weapons and Tactics

At the beginning of the Early Empire, the infantry of the Roman Army already employed a highly standardized panoply that comprised various basic elements: helmet, armour, shield, heavy javelin, short sword and dagger. This had been introduced with the military reform of Gaius Marius, which was implemented shortly before the outbreak of the Roman Civil Wars. Under Augustus, the panoply of the Roman heavy infantrymen was altered in some of its components, but did not change in any radical way: new models of helmet were introduced, the previous *lorica hamata* armour was replaced with the new *lorica segmentata* and the original oval *scutum* was substituted by a new rectangular one. The tactics of the heavy infantry, however, remained more or less the same, since the offensive weapons did not change. The new panoply described above continued to be used by the legionaries until the closing years of the second century. The cavalrymen of the legions were equipped similarly to their foot companions, but continued to use the *lorica hamata* and oval shields also after these were no longer employed by the infantry. The offensive weapons of the horsemen, however, consisted of the spear and long sword instead of the heavy javelin and short sword used by the legionaries. Until the military reform of Augustus, the cavalry and light infantry of the auxiliaries did not have any standard equipment, since each group was usually equipped in its own national fashion. With the development of the *auxilia* system, the many units of non-Roman auxiliaries that were created across the Empire also received a standardized set of equipment that comprised the following elements: helmet, *lorica hamata* armour, oval shield, spear for the cavalrymen or light javelins for the infantrymen, and long sword for the cavalry or short sword for the infantry.

At the beginning of the Imperial Period, the most common model of Roman *galea* ('helmet') was the Montefortino headgear. This was very easy to produce, the Romans having copied it from the Celts, merely introducing some minor adjustments that were needed to transform the Montefortino into the standard helmet of the legions. The name of this model of *galea* derives from the location in Italy where the first helmet of this kind was discovered. The Montefortino had a round shape with a raised central knob – usually surmounted by a coloured plume – and a protruding neck guard generally decorated with incisions. It also had a pair of cheek pieces,

Draconarius from the mid-third century with scale cuirass and metal greaves. (*Photograph and copyright by Conrad Lunar*)

Draconarius from the mid-third century. Note the complex decorations painted on the shield. (*Photograph and copyright by Conrad Lunar*)

Legionary from the years of the Tetrarchy. The practice of painting the shafts of the spears became popular during the third century. (*Photograph by Rachele Lori, copyright by Protectores Domini Nostri and Romars*)

which could be of two main different kinds, according to their shape. The first model reproduced the stylized shape of the cheek and the second was trilobate, consisting of three small discs in a triangular shape. Apparently, trilobate cheek pieces were influenced by the helmets and armour of the Italic peoples, which frequently included this peculiar combination of three discs. However, this style of cheek pieces was gradually abandoned and the other type became dominant among the Roman soldiery. Montefortino helmets had many positive features: they were easy to produce, gave excellent protection to the wearer's face, could be easily decorated in many different ways and were comfortable enough to be worn for a long time. All these characteristics made the Montefortino the most popular helmet of the Roman Republican Period. During the early decades of the Empire, another helmet designed by the Celts and later adopted by the Romans co-existed with the Montefortino: the Coolus helmet. This had a simple round shape, with a ridge running around its base, which on the back enlarged to become a sort of neck guard, similar to the one of the Montefortino helmet. The Coolus helmet also had a couple of cheek pieces that reproduced in a stylized way the human cheek. Being extremely easy and cheap to produce, the Coolus helmet became very popular, especially with the auxiliaries of the Roman Army. Archaeological finds suggest that cheek pieces were only added to Coolus helmets at a later date, as a result of the Montefortino's influence. Consequently, at least initially, the Coolus helmet looked like a simple metal cap. Crest fittings were also only added at a later time.

From the later years of Augustus' long reign onwards, both the Montefortino and Coolus styles started to be replaced by the so-called 'Imperial helmet'. This was produced in two main types – Imperial Gallic and Imperial Italic – which could be in turn divided into further sub-categories corresponding to different versions. The Imperial Gallic helmets, featuring a pair of distinctive embossed eyebrows on the forehead region and tending to be both carefully made and elaborately ornamented, were produced by the skilled Celtic craftsmen of Gaul. The Imperial Italic helmets, lacking the eyebrows and being more roughly made, were produced by less-skilled tradesmen working in Italy. Over time, however, the differences between these two main types started to disappear, as the Imperial Italic helmets also started to be of superior quality. Basically, the Imperial helmet was an improved version of the Montefortino one, with various more advanced features: a sloped neck guard with ribbing at the nape, projecting ear guards, brass trim and decorative bosses. Following the combat experience gained during Trajan's Dacian campaigns, two iron bars riveted crosswise were added across the helmet's skull in order to reinforce it. As an alternative, two thick bronze strips might be riveted to the top of the helmet. From 125, these modifications became permanent features. The Imperial Gallic helmet was produced in three main versions: the original 'G' version, the 'H' version – that had a different style of eyebrows and a more sloping neck guard – and the later 'I' version with a twist-on crest holder instead of the previous slide-on one. The 'I' version, differently from the other two, was made of brass rather than iron. The Imperial Italic helmet was made in four main versions: the original 'D' and 'E' versions, which both had integral brass cross-braces placed flat against the skull but were decorated in different ways; the 'G' version that had brass decorations applied between the crossbars; and the 'H' version, which was made of bronze instead of iron and had a deeper neck guard as well as embossed crossbars. This last version came into use around 180 and had a dome-shaped knob where the cross-braces met at the crown of the head.

The Roman *lorica hamata* was an improved version of the Celtic chainmail, which was designed for the first time during the fifth century BC. This consisted of thousands of small iron rings linked together in order to form a mesh: the rings were interlocked so tightly that they created something similar to a knitted sweater. Depending on its size, a chainmail was made up of some 30,000 metal rings and weighed approximately 10kg. Generally, Roman chainmail was sleeveless but had reinforcement panels for the shoulders: these were attached across the top of the back and held at the front by a bar-and-stud device. A double thong was stretched from the rings attached just above the inner corners of cut-outs to the outer corner of each of the reinforcement panels. The edges of the latter were bound with rawhide, in order to create a raised border. The panels could have angled or rounded-off ends on the chest, and could be made of iron rings like the rest of the armour or alternatively of leather. The Roman *lorica hamata* comprised alternating rows of closed washer-like rings punched from iron sheets and riveted rings made from drawn wire that ran horizontally. These produced very flexible, reliable and strong armour. Each ring had an inner diameter of about 5mm and an outer diameter of about 7mm. The estimated production time for a *lorica hamata* was two months, even with continual slave labour at the state factories. Although labour-intensive to manufacture, this kind of armour, with good maintenance, could be used by a soldier for several decades. Constant friction when worn kept the rings of the *lorica hamata* free of rust, unlike the *lorica segmentata*, which needed continual maintenance to prevent corrosion. In most cases, the *lorica hamata* covered the legs to the knees. Until AD 10–15, all Roman soldiers wore the chainmail; during those years, however, the legionaries replaced it with the new lorica segmentata, the *lorica hamata* continuing to be worn only by the cavalry and the newly organized *auxilia* (both light infantrymen and cavalrymen).

The *lorica segmentata* ('segmented cuirass') was undoubtedly the most advanced kind of armour produced in the Mediterranean region during Antiquity. It consisted

Legionary from the years of the Tetrarchy equipped with Spangenhelm and *manica* of segmented armour. (*Photograph by Rachele Lori, copyright by Protectores Domini Nostri*)

Legionary from the years of the Tetrarchy wearing ridge helmet and scale corselet. (*Photograph by Rachele Lori, copyright by Protectores Domini Nostri*)

of metal strips fashioned into circular bands, which were fastened to internal leather straps. Thanks to archaeological finds, we know that the Roman legionaries fighting at the Battle of Teutoburg (AD 9) employed both the *lorica hamata* and the *lorica segmentata*, so it is possible to say that the segmented armour came into common use during the later years of Augustus' reign. The strips of the *lorica segmentata* were made by overlapping ferrous plates that were riveted to leather straps. The plates were made of soft iron on the inside and rolled mild steel on the outside, in order to be hardened against damage but without becoming brittle. They were hardened by packing organic matter tightly around them and heating them in a forge, in order to transfer carbon from the burnt materials into the surface of the metal. The plates were created by beating out ingots. The strips of each cuirass were arranged horizontally on the body, overlapping downwards; they surrounded the torso in two halves, being fastened at the front and the back. The upper body and the shoulders were protected

Legionary from the years of the Tetrarchy wearing gilded ridge helmet and metal greaves. (*Photograph by Rachele Lori, copyright by Protectores Domini Nostri*)

by additional metal strips. Thanks to its form, a *lorica segmentata* could be stored very compactly; it was possible to separate it into four sections, each of which would collapse on itself, forming a compact mass. The fitments that closed the various plate sections together could be of many different kinds (buckles, lobate hinges, hinged straps, tie-hooks and tie-rings), but were all made of brass. With the progression of time, these started to be replaced by simple rivets, while the lowest plates of the cuirass were replaced by a single and broader plate. Since all the components of the *lorica segmentata* moved in synchronization, this kind of armour was extremely flexible. However, it had two weak points: first, a legionary needed the help of a comrade to wear it; second, it could be severely damaged by rust and thus had to be kept clean. During the Dacian campaigns, in order to protect their arms and legs from the *falx* (sickle) used by most of the enemy warriors, the Roman legionaries started to wear some additional protective elements together with their *lorica segmentata*. To protect their arms they used the *manicae*, arm-guards made of curved and overlapping metal plates that were fastened to leather straps, while to protect their legs they wore simple metal greaves. A single *manica* consisted of one shoulder plate, with about thirty-five strips made of iron or bronze, between ninety and 120 leathering rivets, three or four strips of leather and one padded lining. The metal strips were longer at the top of the arm and all had holes at their lower edge, through which rivets passed from the inside to hold the straps in place. The lower plates of a *manica*, overlapping upwards like all the others, could sometimes be riveted together in order to be stronger.

In addition to *lorica hamata* and *lorica segmentata*, the Roman Army used another two models of armour: *lorica squamata*, or scale armour, that was extremely popular in the eastern regions of the Empire, and the muscle cuirass, that was worn by the Praetorian Guard and the officers. The *lorica squamata* was made from small metal scales sewn to a fabric backing. The individual scales (*squamae*) could be of iron or bronze; the metal was generally not very thick, between 0.5mm and 0.8mm being a common range. Since the scales overlapped in every direction, however, the multiple layers gave good protection. Scales could have rounded, pointed or flat bottoms with the corners clipped off at an angle. They could be flat, slightly domed or have a raised midrib/edge. The scales were wired or laced together in horizontal rows that were then laced or sewn to the backing. Therefore, each scale had between four and twelve holes: two or more at each side for wiring to the next scale in the row, one or two at the top for fastening to the backing and sometimes one or two at the bottom to secure the scales to the backing or to each other. The *squamae* could occasionally be tinned. Scale armour was used by the legionaries but also by the auxiliaries from the eastern provinces, including light infantrymen and archers. The muscle cuirass was a type of body armour cast to fit the wearer's torso and designed to mimic an

Legionary from the years of the Tetrarchy wearing gilded ridge helmet and scale cuirass. (*Photograph by Rachele Lori, copyright by Protectores Domini Nostri*)

idealized human body. It was cast in two pieces, the front and back, which were later hammered together. The cuirasses of officers were highly ornamented with sculpted mythological scenes or similar rich decorations. Usually, the muscle cuirass was used in combination with fringed stripes of leather known as pteruges, which were worn at the armholes and the lower edge of the cuirass.

Until the adoption of the *lorica segmentata*, the Roman legionaries used oval shields, which were copied from those employed by Celtic warriors. These continued to be used by the cavalry and the auxiliaries after the legions adopted the new rectangular shields during the later years of Augustus' reign. The oblong Roman oval shield, or *scutum*, originally had a central spine made of wood and a boss (*umbo*) made of

Legionary from the years of the Tetrarchy wearing gilded ridge helmet and scale armour. (*Photograph by Rachele Lori, copyright by Protectores Domini Nostri*)

Legionary from the years of the Tetrarchy equipped with an extra-long scale corselet. Also the protections for the upper arms are made of scale armour. (*Photograph by Rachele Lori, copyright by Protectores Domini Nostri*)

metal, which were designed to reinforce the whole structure of the shield. Bosses were oblong and could have different shapes. The main body of the shield was made of oak planks, which were chamfered to a thinner section towards the rim. The wooden spine, swelling in the middle, was shaped in order to correspond with a round or oval cut-out in the shield's centre. The strap-type metal boss crossed over the wider section of the spine and was riveted on the external surface of the shield. The surface, either on both sides or only on the front, was entirely covered with leather, which could be painted in various bright colours and have decorations of several kinds. Bosses corresponded to the handle of the shield on the back and thus had a fundamental function in protecting the user's hand. Additional metal binding was frequently attached to the external edges of the shield for reinforcement. The oval shields carried by the light infantry and cavalry were generally a bit smaller than those of the heavy infantry and had a more circular shape. In many cases, those used by the horsemen could be hexagonal in shape rather than oval. The rectangular *scutum* that replaced the oval shield was made from three sheets of wood glued together and covered with canvas, over which there was an external surface of leather. The rectangular shield was convex and had a spindle-shaped boss along its vertical length. With an average weight of 6kg, it was 105.5cm high, 41cm wide and 30cm deep due to its semi-cylindrical nature. Mass production of the scutum was a complex process, with various phases needed to create a shield of good quality. However, its features gave a great tactical advantage to the legionary over most of his opponents. The rectangular shield was light enough to be held in one hand and was large enough to cover the entire body of the soldier, making him very unlikely to be hit by enemy missile weapons. Furthermore, the metal *umbo* in the centre of the shield made it a punching weapon that could be used during hand-to-hand fighting. During the Dacian campaigns, metal edges were added to all Roman shields, since the *falx* of the Dacian warriors could easily penetrate and rip through a scutum. Most of the Roman shields were painted red and decorated with highly standardized devices including eagles' wings, Jupiter's lightning bolts and laurel wreaths.

The first weapon to be used by a Roman legionary in battle was the *pilum*, a heavy javelin specifically designed to kill enemies from long distance or to limit them in the use of their shields. The *pilum*, which was extremely difficult to remove after hitting a shield or a cuirass, had a barbed head and a long narrow socket or shank, made of iron mounted on a wooden haft. The barbs were designed to lodge in enemy shields so that the whole javelin could not be removed. The long iron shank prevented the head from being cut from the shaft. The total weight of a *pilum* was between 1kg and 2kg, meaning the weapon had great penetration power. The shank, due to its peculiar function, was made of softer iron than the hard pyramidal tip; after hitting an enemy shield, it would bend and render the shield useless. The shank of the *pilum* was joined to the wooden shaft by either a socket or a flat tang, which was designed to break when the javelin hit its objective. This meant that when a clash was over, Roman soldiers could recover the wooden shafts from the battlefield and re-use them. In most cases, the *pilum* had a spike on the end of the shaft that made it easier to dig into the ground. Sometimes, the javelin could also be weighted by a lead ball in order to increase its power of penetration. The maximum range of a *pilum* was approximately 35m, although the effective range was some 15–20m. The cavalry and light infantry of the *auxilia* were not equipped with the *pilum*, instead wielding a lighter javelin known as a *verutum*. The shaft of the *verutum* was about 1.1 meters long, much shorter than that of the *pilum*. The tapering point of the *verutum* measured about 13cm and had inferior armour-piercing capabilities compared to that of the heavier *pilum*. The *verutum* had a very simple design, lacking both the shank and the tang of the *pilum*. The legionary cavalry and several mounted units of *auxilia* were equipped with a thrusting spear known as a *hasta*. This had an elongated iron point – with the edges curving inwards from the belly of the blade to its tip – and a wooden shaft. The average length of the *hasta* was 2m, including a butt spike made of iron.

The *gladius* short sword, the most iconic and important weapon of the Roman Army, was not Roman at all. The origins of this weapon are much clearer if we call it by its complete and proper name, the one used by the Romans: *gladius Hispaniensis*. The *gladius* was first designed in Iberia, in the territories of modern Spain and Portugal, as confirmed by various Roman sources but also by archaeological evidence. It was created by the Celtiberians, one of the many warrior peoples inhabiting Iberia during the Iron Age. Differently from the other Iberian communities, the Celtiberians were of mixed descent, being the product of Celtic migrations across the Iberian peninsula. Because of their Celtic heritage, the Celtiberians produced weapons with very innovative techniques. The Romans abandoned their traditional Greek-style swords after the Second Punic War, as a result of their many encounters on the battlefield with the Celtiberian allies and mercenaries of Hannibal. They decided instead to adopt the same short swords as their Celtiberian opponents after realizing that the *gladius* had great tactical potential. According to the latest metallurgical studies conducted on surviving Roman short swords, the *gladius* could be manufactured by two different methods: it could be produced from a single piece of steel or as a composite blade. Swords produced with the first process were created from a single bloom by forging from a temperature of 1,237°C. The carbon content increased from 0.05–0.08 per cent on the back of the sword to 0.35–0.40 per cent on the blade, from which it can be deduced that some form of carburization may have been used. Swords produced via the second process were crafted by the pattern

Lanciarius light infantryman wearing scale corselet. (*Photograph and copyright by Paul Elliott*)

Lanciarius light infantryman wearing Phrygian cap. (*Photograph and copyright by Paul Elliott*)

welding process from five blooms reduced at a temperature of 1,163°C. Five strips of varying carbon content were created. The central core of the sword contained the highest concentration of carbon (0.15–0.25 per cent); on its edges were placed four strips of low-carbon steel (with concentrations of 0.05–0.07 per cent), and the

Lanciarius light infantryman equipped with throwing javelins. (*Photograph and copyright by Varianus/Numerus Brittonum*)

whole thing was welded together by forging on the pattern of hammer blows. Each blow increased the temperature enough to create a friction weld at that spot. The forging operation, the most important part of the whole process, continued until the steel was cold, producing some central annealing. When produced by welding different strips together, the *gladius* had a channel down the centre of the blade, but when made by fashioning a single piece of steel, the blade had a rhomboidal cross-section. The blade of the *gladius* was double-edged for cutting and had a tapered point for stabbing during thrusting. A solid grip on the weapon was provided by a knobbed wooden hilt added on the blade, usually having ridges for the fingers. The hilt, known as the *capulus*, could be decorated in many different ways – the swords of high-ranking officers and Praetorians, for example, usually had a hilt sculpted to resemble the head of an eagle.

The *gladius* continued to be produced for a very long time, in several different designs. According to the traditional subdivision employed by military historians and archaeologists, the various types of *gladius* can be grouped into three main models: Mainz, Fulham and Pompeii (all named after the location where the prototype of each group was found). The original Iberian sword, used from around 200–20 BC, had a slight 'wasp-waist' or 'leaf-blade' curvature that made it quite different from the following models. It was the largest and heaviest type of *gladius* ever produced, with a blade length of 60–68cm and a sword length of 75–85cm. The blade was 5cm wide, with the overall weight of the weapon being some 900g. This earliest form of short sword, still heavily influenced by the original Iberian design, was used for a very long period compared with its successors. The Roman city of Mainz in the Rhineland was founded, like many other Roman centres, as a permanent military camp named Moguntiacum around 13 BC. The original military camp soon became an important centre for the production of swords and other military equipment. With the transformation of the camp into a city, the manufacture of swords became even more significant, leading to the creation of a new kind of *gladius* which is commonly known as the Mainz *gladius*. The Mainz *gladius* kept the curvature of the previous model, but shortened and widened the blade. In addition, it modified the original point into a triangular one, specifically designed to thrust. The geographical diffusion of this model was limited to the border garrisons serving on the northern frontiers, differently from the less battle-effective Pompeii version that came into use in other areas of the Empire. The short swords produced at Mainz during the early Imperial Period were used by the legions serving in the north, along the *limes*, but not only there, as large numbers of these weapons were exported and sold extensively outside the boundaries of the Roman Empire. Apparently, various veteran ex-legionaries who had served on the frontiers used their discharge bonus upon retirement to set up

Lanciarius light infantryman armed with throwing javelins. (*Photograph and copyright by Varianus/Numerus Brittonum*)

Unarmoured legionary wearing a model of ridge helmet with metal crest. (*Photograph by Rachele Lori, copyright by Protectores Domini Nostri*)

Lanciarius light infantryman equipped with round shield. (*Photograph by Rachele Lori, copyright by Protectores Domini Nostri and Romars*)

businesses as manufacturers and dealers of arms. The Mainz variety of the *gladius* was characterized by a slight waist running the length of the blade. The average Mainz *gladius* had a blade length of 50–55cm and an overall length of 65–70cm. The blade was 7cm wide, and the overall weight of the weapon was around 800g. The Fulham *gladius* derives its name from the fact that the original prototype of this model was dredged from the River Thames near Fulham in west London. The Fulham *gladius* had a slightly narrower blade than the *Mainz* variety, but the main peculiarity of this model was its triangular tip. It had a blade length of 50–55cm, sword length of 65–70cm, the blade was 6cm wide and the overall weight of the weapon was 700g. The Pompeii *gladius*, which was without doubt the most popular of the three models that started to be produced after the original Iberian one, had parallel cutting edges and a triangular tip. From a structural point of view, the Pompeii model eliminated the curvature, lengthened the blade and diminished the point. It was the shortest model of *gladius*: this was the result of battle experience in the Roman Civil Wars, during which Roman armies fought against each other in large pitched battles. In these cruel internal conflicts, hand-to-hand tactics were employed by both sides, meaning the traditional Roman military superiority lost most of its usual efficiency. Having to fight against enemies equipped exactly like themselves, with heavy cuirasses and shields, the Romans had to develop a lighter and shorter version of their sword, designed to thrust mainly with the point and in very tight spaces. The average Pompeii *gladius* had a blade length of 45–50cm, a sword length of 60–65cm, the blade was 5cm wide and the overall weight of the weapon was 700g. Over the years, with the end of the Roman Civil Wars, the Pompeii *gladius* gradually became longer. All types of *gladius* were carried in a scabbard mounted on a belt or, more frequently, on a shoulder strap. It was worn on the left side of the soldier's body, meaning each legionary had to reach across his body to draw it. Centurions, to differentiate themselves from their soldiers, wore the *gladius* on the right side of their body.

The Roman cavalrymen, be they legionary horsemen or auxiliaries, were not armed with the *gladius* of the foot legionaries/auxiliaries but with a longer *spatha* or cavalry sword. This, originally designed by the Celts, had a distinctive elongated leaf shape, being double-edged and having a square-kink or shallow 'V' point (with the sides drawn at an angle of 45 degrees to the axis of the blade). The tang of the *spatha* – the internal part of the handle, made of metal but covered with organic material – swelled sharply to a point of greatest width that was just below its centre. The ricasso – the unsharpened length of blade just above the handle of the sword – was very short and had a notch that varied greatly in depth. Sword handles were made of wood or leather and generally had the form of an 'X', continuing the pattern of the famous Celtic antennae swords. The handle was completed by a pommel, which was

Archers from the years of the Tetrarchy. The figure on the left is armoured. (*Photograph by Rachele Lori, copyright by Protectores Domini Nostri*)

Archer from the years of the Tetrarchy. He is armed with a mace in addition to his bow. (*Photograph by Rachele Lori, copyright by Protectores Domini Nostri*)

Legionary from the mid-third century wearing the clothes used while off-duty. Note the wide-brimmed sun hat made of wicker. (*Photograph and copyright by Varianus/Numerus Brittonum*)

connected to the tang by a rivet-hole. The blade measured between 60cm and 90cm in length and was made of iron or steel. In most cases, blades had a broad neck, with the greatest width being usually low down towards the point. Over time, points started to be increasingly rounded, this shape confirming that these cavalry weapons were used for slashing and not for thrusting. Roman long swords were transported in iron scabbards, richly decorated with incisions and/or bosses. The scabbards reproduced the general shape of the blade and were constructed from two plates, with the front one, slightly wider than the back one, folded over the latter along the sides. Each scabbard was reinforced by a decorated band around the top and a sculpted chip at the bottom. Scabbards were generally suspended on the right hip from a leather belt and attached to the waistbelt by means of a metal loop on the back of the scabbard. The *spatha* was produced in four slightly different models during the third century, known as the Illerup Adal, Ring Pommel, Vimose and Thorsberg types (as shown in the photographs accompanying this text).

Each Roman soldier, in addition to the *gladius* or *spatha*, also had a short dagger known as a *pugio*. This sidearm was intended as an auxiliary stabbing weapon, specifically designed for hand-to-hand combat. Like the *gladius* short sword, it was designed by the Celtiberians and later adopted by the Romans. The *pugio* had a large leaf-shaped blade, which usually narrowed from the shoulders to run parallel for about half of its length before narrowing to a sharp point. A midrib ran close to the length of each side, either standing out from the face or sunken by grooves on either side. The tang was wide and flat; the grip was riveted through it, as well as through the shoulders of the blade. The pommel was originally round, but it soon assumed a trapezoidal shape. Each *pugio* was carried in its own sheath made of soft metal, which had four suspension rings and a bulbous terminal expansion that was pierced by a large rivet. Over time, the original design of the *pugio* was slightly modified, with a rod tang introduced and the hilt no longer being riveted through the tang as it started to be secured only at the shoulders of the blade. The hilt continued to consist of two layers of horn or wood. The *pugio* could be 18–28cm long and was 5cm wide. The sheath could be heavily decorated, especially for the daggers used by officers or Praetorians.

Each legionary also carried a *sarcina*, a military pack, which contained several important pieces of equipment. Since the days of Gaius Marius' military reforms, the Roman heavy infantrymen had become famous for their capabilities as combat engineers, being able to build fortified camps or military infrastructure such as bridges in every condition and on every kind of terrain. The *sarcina* consisted of a *furca*, or carrying pole, to which the following objects were suspended: a *loculus* (leather satchel), a waterskin, a small bag with rations of food for fifteen days, a *patera* shallow dish, cooking pot and skewer, some pointed stakes (*sudis*) used to construct fortifications, a *dolabra* pickaxe and *ligo* mattock, one *falx* to clear undergrowth, a *batillum* shovel and a basket for hauling earth. The equipment of each legionary was completed by the *cingulum militare*, a military belt which had a very important symbolic significance: wearing it meant being part of the army, since civilians used different models of belt. Depriving a soldier of his military belt meant expelling him from the army. The *cingulum militare* was a leather belt decorated with iron fittings and consisting of the following components: the *baltea* (hanging band), *bulla* (rivets that were applied on the *baltea*), *pensilium* (pendants that were applied at the end of the belt's straps), *lamna* (disc that was applied at the end of each apron strip holding the *pensilium*) and *fibula* (belt buckle).

It is not strictly correct to speak of 'uniforms' for the armies of Antiquity, but it is possible to say that the Roman Army of the Early and Middle Empire had a high level of standardization in its dress, mainly due to the fact that state factories produced military clothing, equipment and weapons for all the soldiers of the Empire. The main garment worn by the Roman soldier was the tunic, which was generally short-sleeved but was also produced in a version with long sleeves known as a *tunica manicata*, which was commonly used in the northern provinces of the Empire. The Roman tunic, made of wool for cold months and of linen for hot months, was a simple T-shaped garment reaching the knees of its wearer. It was not made of cut-and-sewn parts, but was woven entire in a single piece, each *tunica* being folded in half and sewn up the sides, with a neck slit cut in the centre. The standard colours for military tunics were red and white, with the latter being extremely popular in the hottest regions of the eastern provinces. The cloak was the second most important garment for the Roman soldier. Made of wool and containing natural oil in order to repel water, it could be of two different kinds, *sagum* and *paenula*. The *sagum*, typical of the Italian peninsula, was rectangular in shape, whereas the *paenula* was rounded and had a hood (*cuculia*) for protection of the head from the elements. All cloaks reached the hips and were double-folded so that they did not fall beneath the knees when worn. The cloaks of officers were quite different from those of the ordinary soldiers, being longer (ankle-length) and trapezoidal in shape. All cloaks were pinned up on the right shoulder by a large brooch (*fibula*), which could be richly decorated if worn by a senior officer. Although trousers, known as *bracae*, were considered by most well-educated Roman citizens to be a 'barbarian' garment, the legions garrisoning the northern frontiers widely adopted them from the early days of the Empire. All legionaries wore a *focale*, or scarf, around their neck, being made of wool for the winter and linen for the summer. The *focale* protected the neck from chafing by the armour and was loosely knotted at the front. For footwear, all Roman soldiers wore

Legionary from the mid-third century wearing the clothes used while off-duty. Note the wide-brimmed sun hat. (*Photograph and copyright by Varianus/ Numerus Brittonum*)

the open sandals known as *caligae* (a name deriving from Emperor Caligula, who loved wearing them). These were made from leather and were laced up the centre of the foot to the top of the ankle. Iron hobnails were hammered into their soles for added strength.

Rank distinctions existed in the Roman Army. A centurion, for example, was distinguished as follows: he carried a *gladius* on his left and a *pugio* on his right (the opposite side to the common legionaries); he had a peculiar transverse crest on his helmet; he always carried a vine stick (*vitis*) that was used to mete out punishments; and he had sculpted discs known as *phalerae* (made of gold, silver or bronze) on the front of his armour. The *phalerae* were military medals, usually awarded for distinguished conduct in action. They were aligned in three horizontal rows and were attached to orthogonal leather straps in order to form a single breastplate that was worn on the chest. An *Optio*, meanwhile, was easy to recognize as he had plumes of horse hair or feathers applied on either side of his helmet (the standard crest being in the middle). The *Optio* also carried a long wooden staff (*hastile*) that was as tall as the *Optio* himself and had a round metal pommel at its end. Like the vine stick of centurions, this was used to mete out punishments. All standard-bearers carried a peculiar

Officer from the years of the Tetrarchy wearing *pileus pannonicus* made of fur. (*Photograph by Rachele Lori, copyright by Protectores Domini Nostri*)

Legionary cavalryman from the early third century, equipped with Heddernheim cavalry helmet. (*Photograph by Simone Ochsner (https://www.simoneochsner.com), copyright by Legio XI CPF*)

small round shield and wore an animal fur over their shoulders and head: a wolf for the *Vexillifer*, a bear for the *Signifer* and a lion or tiger for the *Aquilifer*. The fur was worn as a cape, with the head of the animal strapped to the helmet of the standard-bearer. In many cases the standard-bearers, being veterans, wore on the chest the same *phalerae* as the centurions. The military musicians of a legion were usually equipped like the standard-bearers, carrying a small round shield and wearing a fur over their shoulders and head.

The general appearance of the Roman soldiers during the Crisis of the Third Century progressively became quite different from that reproduced on Trajan's Column, with the traditional image of the legionaries equipped with segmented armour and the auxiliaries with chainmail no longer being in existence by 284. The Crisis of the Third Century saw a progressive collapse of the Roman central administration, which had important consequences for the organization of Rome's military forces but also for the supply system that had until then worked very well. During those years of anarchy, the regular soldiers started to receive new clothes and weapons on a much less regular basis, resulting in a partial loss of the high level of standardization that had previously been achieved. Diocletian and Constantine did what they could to rebuild the supply system of the armies, the *Notitia Dignitatum*, for example, listing a total of thirty-five state *fabricae* (factories) located across the Empire and producing military materials. The previous system was thus at least partially restored. Each state factory was under the control of a Procurator and was even protected by a small military garrison. The cost of the production of military materials were still paid by the central administration through a specific tax. After the definitive division of the Empire in 395, however, the different levels of uniformity between the military units of the Western Empire and those of the Eastern Empire became apparent. The tax revenues of the eastern regions were almost double those in the West, with the result that soldiers from Gaul or Italy started to be equipped with materials of inferior quality. Generally speaking, the clothing and equipment of the *comitatenses* showed a higher degree of uniformity than those of the *limitanei*, the latter mostly purchasing their clothing from small local stores attached to their frontier forts. As had been the case for the *auxilia* of previous centuries, the weapons and equipment of the *limitanei* were strongly influenced by the military fashions of the peoples living along the Roman frontiers (both inside and outside the Empire).

During the Middle Empire, the main garment worn by the Roman soldiers was still the tunic, although even this had changed a lot since the time of Augustus. The Arch of Severus, which was completed at the beginning of the third century, still shows Roman soldiers with the traditional kind of tunic worn during the Early Empire, but this was soon changed by the following years of crisis. Indeed, the central

Legionary cavalryman from the early third century. Note the complexity of the decorations painted on the shield, which is a parade one. The visor and the facial mask of the helmet could be removed if needed. The panoply shown here was also used during the Roman cavalry's sports game known as '*hippika gymnasia*'. (*Photograph by Simone Ochsner (https://www.simoneochsner.com), copyright by Legio XI CPF*)

Legionary cavalryman from the early third century. He is wearing a leather corselet with two decorative plates of metal on the chest. Note the rich decorations of the massive greaves. (*Photograph by Simone Ochsner (https://www.simoneochsner.com), copyright by Legio XI CPF*)

decades of the third century saw the widespread introduction of the new long-sleeved tunic, which was copied from the contemporary dress of Germanic warriors, who had used it on a large scale for a long time. The Roman terminology for this garment was *tunica manicata* and it was still generally made of wool during the winter and linen during the summer. Apart from its long sleeves, the *tunica manicata* generally retained the same characteristics as the short-sleeved version. However, it was now generally quite decorated, with woven or attached strips (*clavi*) and circular roundels of cloth (*orbiculi*). Both these decorative elements usually comprised geometrical patterns and stylized plant motifs, but could include also human or animal figures. The *clavi* were narrow vertical bands of cloth running from the top to the bottom of the tunic, while the *orbiculi* were patterned roundels on the shoulders and skirt of the tunic. The standard colours for military tunics were still red and white, with white versions being particularly prevalent in the hottest eastern regions.

After the tunic, the cloak remained the second most important garment of the Roman soldier. Hip-length and semi-circular, the *sagum* could be decorated with fringes or with applied decorations of the same kind worn on the tunic. The cloaks of officers were quite different from those of soldiers: they were longer (ankle-

The Hallaton helmet, produced during AD 30–50. (*Photograph by Simone Ochsner (https://www.simoneochsner.com), copyright by Legio XI CPF*)

length) and trapezoid in shape, and were not decorated with *clavi* or *orbiculi* but with large embroidered panels having the form of a square, known as *tabula*. Cloaks were still pinned up on the right shoulder by a large brooch, which could be richly decorated for senior officer. Materials and colours were more or less the same as the tunics: wool for winter and linen for summer, red and white being the most popular colours. During the late third century, Roman military belts started to show a strong Germanic influence, especially in their decoration. Bronze or iron fittings became increasingly popular, both cast and chiselled, being embellished with punched or carved work. By the time of Diocletian and Constantine, most of the soldiers wore *bracae*. Meanwhile, the traditional open sandals (*caligae*) had completely disappeared by the end of the third century, being replaced by new front-fastening short boots with integrally cut laces. For headgear, when not wearing the helmet, soldiers on active service usually had a particular kind of cap known as the *pileus pannonicus*. These could be made from felt or napped wool, and could be smooth or shaggy, low or tall. The cylindrical shape made this kind of cap very comfortable and easy to produce. It appears it started to be used during the early third century, also being adopted as a sign of distinction by civilian dignitaries. This pillbox cap was usually covered with fur in the northern regions of the Empire.

During the Crisis of the Third Century, all the various kinds of helmet traditionally worn by the Roman soldiers were progressively abandoned and replaced with new ones. This happened mainly for two reasons. Firstly, the previous models of helmet were elaborate and thus required particular skills in order to be made. In an age of anarchy, characterized by the collapse of the military supply system, mass production required helmets with simpler designs. The second factor behind the change of headgear was the influence of the Sarmatians and Sassanids, the Romans having learned from their new enemies how to produce cheaper and simpler models of helmet. These are commonly divided into two major families: ridge helmets and spangenhelms. Ridge helmets were characterized by a bowl made of two or four parts, united by a longitudinal ridge. This method of construction was copied by the Romans from Sassanid helmets. The earliest confirmed example of a Roman ridge helmet is the Richborough helmet, which dates back to 280. Differently from earlier Roman helmets, the skull of the ridge helmet was constructed from more than one element. Roman ridge helmets can be classified by two types of skull construction: bipartite and quadripartite, respectively referred to as Intercisa-type and Berkasovo-type. Bipartite helmets had a two-part bowl united by a central ridge (running from front to back), and had small cheekpieces. They had no base-ring running around the rim of the bowl. Some surviving examples of bipartite helmets have metal crests, such as those known as Intercisa-IV and River Maas. Quadripartite helmets featured a four-piece bowl connected by a central ridge, with two plates (connected by a reinforcing band) on each side of the ridge and a base-ring uniting the elements of the skull on the rim of the helmet. This type of helmet had large cheekpieces. Some surviving examples of the quadripartite construction also have a nasal. It is believed that the cheekpieces were attached to the skull of the helmet by a liner, and that a separate neck guard was frequently attached to the back of the helmet by flexible leather straps. The bipartite construction or Intercisa-type had ear-holes, while the

Example of a Niederbieber helmet, one of the models that were most used during the third century. (*Photograph and copyright by Varianus/Numerus Brittonum*)

quadripartite construction or Berkasovo-type did not. Some scholars, bearing in mind that earlier Roman cavalry helmets usually had cheek guards with a section covering the ears, have suggested that the Intercisa-type was used by infantrymen and the Berkasovo-type by cavalrymen. This interesting hypothesis is supported by an important archaeological find, the Deurne helmet, which has an inscription relating to a cavalry unit of the *Equites stablesiani*. It should be borne in mind, however, that both types of ridge helmet are depicted in several works of art as being worn by both infantry and cavalry, so it is highly probable that they were used interchangeably.

Example of a Niederbieber helmet of the Frankfurt type, one of the models that were most used during the third century. (*Photograph and copyright by Varianus/Numerus Brittonum*)

The spangenhelm arrived in Europe by way of what is now southern Russia and Ukraine, being spread by nomadic peoples including the Sarmatians, who lived on the Eurasian steppes. However, it should be noted that the term 'spangenhelm' is clearly of Germanic origin: 'spangen' refers to the metal strips that form the framework of the helmet, while 'helm' simply means 'helmet'. The characteristic metal strips of a spangenhelm connected between three and six steel or bronze plates. These

Example of a Niedermörmter helmet, one of the models that were most used during the third century. (*Photograph and copyright by Varianus/Numerus Brittonum*)

made up a framework that had a conical design, which curved with the shape of the head and culminated in a point. The front of the helmet generally included a nasal. Spangenhelms could also incorporate chainmail as neck protection, forming a sort of aventail at the back. Some surviving examples also include eye protection, with a shape that resembles modern eyeglass frames, whereas others include a full face mask. Older spangenhelms often had cheek flaps made from metal or leather. The spangenhelm offered an effective protection for the head that was relatively easy to produce. Both the new families of Roman helmets could have heavy decoration, especially if a helmet belonged to a rich officer. Many surviving examples of helmets from the Middle Empire have evidence of decorative silvering or are covered by costly silver-gilt sheathing. The job of decorating helmets was entrusted to special artisans working in the state factories, who were known as *barbaricarii*. Some helmets of senior officers, like the Berkasovo-I helmet, were decorated with several glass gems (located on the bowl, the cheekpieces and the neckguard).

With the *Constitutio Antoniana* of 212, as we have already seen, the main social distinction between legionaries and auxiliaries was suddenly cancelled: all the inhabitants of the Empire became Roman citizens, meaning there was no more sense in the differences of equipment between the two types of soldiers. Until the Crisis of the Third Century, legionaries were equipped with the segmented armour known as *lorica segmentata*, while the auxiliaries wore the traditional *lorica hamata* chainmail armour. As it did with helmets, the Crisis of the Third Century represented a real turning point for Roman armour: *lorica segmentata*, being quite complex and difficult to produce, was progressively abandoned during the turbulent decades following 235. The traditional segmented armour is still reproduced on the Arch of Constantine erected in 315, but we should bear in mind that these depictions come from an earlier monument erected by Marcus Aurelius that were later incorporated into the Arch of Constantine. The *lorica segmentata* thus probably disappeared around the mid-third century. It was replaced by the *lorica hamata* formerly worn only by the auxiliaries, which now started to be used by all units. The basic features of the Roman chainmail remained more or less the same. It comprised alternating rows of closed washer-like rings punched from iron sheets and rows of riveted rings made from drawn wire that ran horizontally, which produced armour that was very flexible but strong. Each ring still had an inner diameter of about 5mm and an outer diameter of about 7mm. Up to 30,000 rings went into each *lorica hamata*, which took about two months to manufacture. Most *lorica hamata* covered the legs to the knees, and its sleeves could be long or short. *Lorica hamata*, however, was not the only kind of armour worn by the soldiers of the Middle Empire, there being three other types, which had very different origins and features.

The first of these was the already popular *lorica squamata* (scale armour) that was made from small metal scales sewn to a fabric backing. The individual *squamae* could be iron or bronze, and as we have seen, were not very thick, but as the scales overlapped in every direction, the multiple layers offered wearers good protection. Scale armour had been used in Asia since the days of the Sumerians, so *lorica squamata* was extremely popular in the eastern provinces of the Empire. Mail armour, instead,

Detail of the decorated gilded plates worn on the chest of a scale corselet. (*Photograph by Rachele Lori, copyright by Protectores Domini Nostri*)

was dominant in the western provinces. During the Crisis of the Third Century, however, scale armour started to be widely used by legionary units in the western part of the Empire. The second kind of armour used by the Roman Army of the Middle Empire was lamellar, but this was employed on a much smaller scale than those discussed previously because it was imported into Europe from the steppes of Eurasia only during the very last decades of the Middle Empire. It was also mostly used by mercenary *buccellarii* or by *foederati*, and not by regular Roman soldiers. Lamellar armour consisted of small platelets, commonly known as *lamellae*, which were punched and laced together to be assembled in several horizontal rows. The *lamellae* could be made of metal, leather or horn; metal ones could be lacquered in order to resist corrosion or as decoration. Differently from scale armour, lamellar armour was not attached to a cloth or leather backing. The third kind of armour worn in addition to the *lorica hamata* was the muscle cuirass. During the Middle Empire, simpler versions of this were usually preferred, with the sculpted anatomy reduced to

Nice example of gilded ridge helmet. (*Photograph by Rachele Lori, copyright by Protectores Domini Nostri and Legio Palatina Ioviani Seniores*)

an abstract design. The cuirasses of senior officers were often highly ornamented with sculpted mythological scenes or similar kinds of rich decoration. The muscle cuirass was usually used in combination with fringed leather pteruges at the armholes and the lower edge of the cuirass. The 'muscle cuirass' was mostly worn by officers, being considered a sign of their rank. The Columns of Arcadius and Theodosius, as well as other examples from Roman art, suggest that the muscle cuirass was in common use among the *Scholae Palatinae* and *Protectores* units. Since the days of the Praetorians,

Nice example of crested ridge helmet. (*Photograph by Rachele Lori, copyright by Protectores Domini Nostri and Legio Palatina Ioviani Seniores*)

Nice example of Spangenhelm. (*Photograph by Rachele Lori, copyright by Protectores Domini Nostri*)

the muscle cuirass had been the favoured personal protection worn by all Roman guard units, and this general trend apparently continued during the later centuries. Generally speaking, neither light infantry nor light cavalry wore armour, but the heavy cavalry used the same kinds of armour described above for the foot troops.

During the course of the third century, the traditional *scutum* used by legionaries disappeared along with the *lorica segmentata*. Mass production of the scutum was a quite complex process, with various phases needed to create a shield of good quality. As with other components of the standard Roman military equipment, the production of the rectangular shield proved unsustainable for the manufacturing structures and supply system of the Empire during the chaotic years of the third century. As a result, by the time of Constantine, almost all Roman military units had adopted a new and simpler model of oval shield. This was carried by light infantry and cavalry too, although with some possible variations in dimensions (the shields of light infantry and cavalry were generally a bit smaller than those of the heavy infantry, having a

Legionary equipped with Spangenhelm and gilded muscle cuirass. (*Photograph by Rachele Lori, copyright by Protectores Domini Nostri*)

more circular shape). Unlike the *scutum*, the new oval shield was not convex but could be either dished (bowl-shaped) or flat. This difference in shape obviously had some consequences on tactics, but it seems that the new model of shield was still sometimes used to form the traditional defensive formations. The oval/round shield was much larger than the *scutum* and was constructed in a different way, being made of solid planks instead of plywood and supported by a double grip (at the elbow and hand), whereas the *scutum* had just a single central grip). The general features of the new type of shield made it much easier to produce than the previous rectangular one. In many aspects, it was the direct heir of the oval shield used by the *auxilia* for their entire history. The new oval shield was about 110cm high and 90cm wide, being constructed using 1cm thick wood planks, and was covered and bound with leather. A hollow iron or bronze boss covered the central hand grip. The *scutum* had usually been painted with standard decorations, which were almost identical for each legion, and the shields of the auxiliaries were also generally painted with standard motifs. It was not possible, however, to distinguish one unit from another thanks to the decorations reproduced on its shields. After the reforms of Diocletian and Constantine, the new oval/round shield started to be painted with different motifs according to the various units. Each unit of the Roman Army was given its distinctive shield emblem. This was quite revolutionary and modern for the time, since it reflected a very precise unit organization. To help reconstruct the shield emblems of the Late Roman Army, historians are aided by the exceptional primary source of the *Notitia Dignitatum*. Nevertheless, the shield emblems of some categories of troops have not survived in this source and thus have been lost. It is important to note, however, that the units of *limitanei* listed in the *Notitia Dignitatum* have no corresponding shield emblem: some scholars think that these were lost, but it is more reasonable to suppose that the frontier troops had no distinctive unit emblems. Only the *comitatenses*, the mobile units, had distinctive emblems painted on their shields, because they could operate in various regions of the Empire and thus needed to be identified by commanders of the field armies. The *limitanei*, instead, served only in their home territories and were not employed in other areas of the Empire: for this reason, they had no need to be identified via their shield emblems. The various unit emblems reproduced in the *Notitia Dignitatum* are quite complex, but it is clear that the motifs painted on their shields by individual soldiers were simplified and stylized. In most cases, each soldier might have managed to paint just a quick and simple copy of his unit's emblem. The variety of symbols and colours, mixed in various combinations, appearing in the shield emblems of the *Notitia Dignitatum* is simply astonishing, so it is very hard to believe that each soldier of every unit respected such a complex system of emblems.

While the Roman legionaries of the Early Empire could count on excellent defensive equipment, comprising the *lorica segmentata* and *scutum*, the real strength of the Roman heavy infantryman was his offensive weapons: the *pilum* (heavy javelin), *gladius* (short sword) and *pugio* (dagger). The first weapon to be used in battle was the *pilum*, a javelin specifically designed to kill enemies from distance or to limit them in the use of their shields (the *pilum* being extremely difficult to remove after hitting a shield or cuirass). Once the enemy ranks had been shattered by the initial rain of javelins, the legionaries drew their short swords and charged their opponents. According to Roman tactical doctrine, emphasis was placed on using the *scutum* to provide maximum body coverage, while the *gladius* was used to attack with devastating thrusts and short cuts. Employing such tactics, the Romans were for several centuries able to defeat any enemy infantry they faced. This kind of warfare also had the benefit of limiting the number of casualties suffered by the imperial troops. Using their swords to thrust in the few spaces created between the shields of their close formations, the legionaries were rarely exposed to the offensive weapons of their enemies, who had very few opportunities to manoeuvre. Like the *gladius*, the *pugio* dagger was a short stabbing weapon. It was used as a secondary armament during the most intense hand-to-hand fighting, especially when space for movements became very limited or when the *gladius* could not be used for any reason.

Soon after the beginning of the Crisis of the Third Century, the traditional Roman system of offensive weapons started to be completely modified. The short *gladius* stabbing sword designed for close-quarter fighting was gradually phased out by the new *spathe*, a long sword that had originally been used only by cavalrymen. At the same time, the *pugio* dagger was progressively abandoned as a secondary hand weapon, being replaced by simpler short and single-edged knives (which, unlike the *pugio*, were not carried by all soldiers on a regular basis). The adoption of the *spatha* as the universal sword of the Roman Army was one of many important military changes during the third century, with Roman tactics being slowly transformed as a result of the new threats faced by the Empire. The new prominence of cavalry had transformed the Roman Army into a highly mobile force mostly composed of heavy armoured horsemen and mounted archers. As a result of the cavalry's predominance on the fields of battle, the close infantry formations designed for hand-to-hand fighting were gradually abandoned and the long slashing swords of the cavalry also started to be employed by the infantry. The cavalry's long sword had been copied by the Romans from the Celts, similarly to how the *gladius* had been copied from the Iberian warriors of Hannibal who fought against Rome during the Punic Wars. The *spatha* was introduced into the Roman Army by the auxiliary cavalry units formed from Celtic recruits of Gallic origin. Generally speaking, the units of *auxilia* retained their own weapons and fighting traditions after joining the Roman Army. The Celtic cavalry, famous for their excellent long swords, obviously continued to use these superior weapons. Consequently, from the early days of the Empire, the Celtic *spatha* started to be used by the majority of the Roman cavalry units. This straight and long sword had a blade length between 0.75m and 1m (much longer than a *gladius*). After being used for several decades only by the cavalry, during the third century the *spatha* started to also be carried by heavy infantry soldiers. The *gladius* was initially relegated to use as a light infantry weapon, but was soon totally abandoned. The spatha now

Ridge helmet with special protections for the ears. (*Photograph by Rachele Lori, copyright by Protectores Domini Nostri and Legio Palatina Ioviani Seniores*)

started to be produced in two different versions, one for the infantry and another for the cavalry. The infantry version had a long point, while that for the cavalry had a rounded tip that prevented accidental stabbing of the horseman's foot.

Since the days of the Early Empire, the Roman legionary's fighting style had been defined by two weapons: the *pilum* and the *gladius*. The *pilum*, with a long iron shank for punching through enemy shields and armour, was usually thrown

Pileus pannonicus made of felt. Such headgear was extremely popular during cold months, especially among the legions stationed in continental Europe. (*Photograph by Rachele Lori, copyright by Protectores Domini Nostri*)

at close range. It gave the Roman legionaries a great tactical advantage over their enemies, because it was a very effective light infantry weapon in the hands of heavily armoured infantrymen. No other soldier in the Mediterranean world had such excellent defensive equipment used together with a deadly missile weapon like the *pilum*. The traditional heavy *pilum* was progressively abandoned from around 250, during the Crisis of the Third Century. Differently from other pieces of traditional Roman military equipment, it was not replaced by a single new weapon. The Roman infantrymen of the Middle Empire could now use two different kinds of javelin to break up the ranks of the enemy. Both these new missile weapons – the *spiculum* (heavy javelin) and *verutum* (light javelin) – had butt-spikes that protected the base of the shaft from rot and could also be used for thrusting. The *spiculum* replaced the *pilum* as the heavy infantryman's main throwing weapon around. Scholars have conjectured that this new weapon could have resulted from the gradual combination of the *pilum* with the standard Germanic javelin, known as the angon. The latter was quite similar to (and probably derived from) the *pilum*, having a barbed head and a long narrow socket or shank, made of iron mounted on a wooden haft. The barbs were designed to lodge into enemy shields so that it could not be removed. The long iron shank prevented the head from being cut from the shaft. The angon was likely designed like the *pilum* to disable enemy shields, leaving opponents vulnerable and disrupting enemy formations. The shaft may have been decorated, and iron or bronze rings were sometimes fitted onto it, which may have marked the centre of balance and thus the best place to hold the weapon. The Roman *spiculum* had all the main features of the Germanic angon, the main difference between it and the *pilum* being the length of the thin point, the new javelin tending to have a much shorter point than the previous version. In general, the *spiculum* was significantly shorter than its predecessor, perhaps being some 190cm long. Its exact design is not yet fully known, because there were many variants, but what is known is that it had a medium iron shank attached to the head. In addition to the *spiculum*, the soldiers of the Middle Empire could also employ a lighter and longer-ranged javelin known as a *verutum*. This was the main missile weapon of light infantry and cavalry, but could also be used by heavy infantrymen in combination with the *spiculum*. The *verutum* had been used for skirmishing purposes since the days of the Republic, being issued to the light infantrymen of the legions. During the Crisis of the Third Century, however, it started to be employed by all kinds of units. The shaft of the *verutum* was about 1.1m long, much shorter than that of the old *pilum*. The point measured about 13cm, while each *verutum* had either an iron shank like the *spiculum* or a tapering metal head. The *verutum* had a narrow armour-piercing head similar to that of the heavy javelins, though obviously with inferior capabilities of penetration. Over time, it seems that the *spiculum* started to be used on a lesser scale, apparently being replaced by the *verutum* as the main javelin for heavy infantry.

In addition to the above, the soldiers of the Middle Empire could use some very innovative missile weapons known as *plumbatae* or *martiobarbuli*. These were lead-weighted darts, usually carried by the heavy infantrymen on the back of their oval shields. Initially introduced in Diocletian's newly created elite *legiones palatinae* of the *Ioviani* and *Herculian*, these highly effective darts rapidly spread to the other legions and *auxilia palatina* units. They were in essence a short arrow with a weight attached

Detail of the military decorations known as *phalerae* that were worn by a centurion across the chest. (*Photograph by Simone Ochsner (https://www.simoneochsner.com), copyright by Legio XI CPF*)

to the shaft. The term *plumbatae* etymologically contains the word *plumbum*, which meant 'lead' in Latin, so it is possible to translate *plumbatae* as 'lead-weighted darts'. The term *martiobarbuli* is most likely an assimilation of the words *martio* and *barbuli*, and can thus be translated as 'little barbs of Mars'. This second etymology clearly implies a barbed head. An important archaeological find from Wroxeter in Shropshire finally confirmed the general picture scholars had of the *plumbatae*: fletched darts with an iron head, weighted with lead. They had little penetrating power and a very short range (just 30m), but their barbed heads made them very efficient for crippling an enemy or a horse without armour. Each heavy infantryman carried six of these weapons clipped to the back of his shield.

Example of *subarmalis*, the protective padded garment worn by legionaries under their metal cuirasses. (*Photograph and copyright by Varianus/Numerus Brittonum*)

Since the reforms of Gaius Marius, the Roman heavy infantry had abandoned the spear as one of its main offensive weapons, with all legionaries starting to be equipped with the *pilum*. The infantry of the auxiliary units, however, did not receive the *pilum* and continued to employ the normal model of spear (*lancea*). As we have seen, the legionaries started to abandon the *pilum* around 250, it being replaced by the new *spiculum* as heavy javelin, but a good number of legionaries started to adopt the *lancea* of the auxiliaries as their main weapon. As a result, after a gap of several centuries, the Roman heavy infantry was again equipped with spears on a large scale. The *lancea* of the Middle Empire was slightly different from that previously wielded by the auxiliaries: it was light enough to be thrown just prior to contact but could be retained for use in hand-to-hand combat. Apparently, judging from the primary sources that are available, the heavy infantrymen of the Middle Empire preferred using the *lancea* for close fighting. Indeed, each soldier usually also carried two or three *spicula* that were to be used as throwing weapons. Very often, however, a legionary could be armed only with a *lancea*, although the soldiers of the *auxilia palatina*, being elite light infantrymen, obviously were mostly equipped with the *spiculum* or the *verutum* (the latter being the standard weapon of all light infantry and cavalry). The wood shaft of the *lancea* could be painted in various colours to prevent rot, but this was not always possible under campaign conditions.

While this was the situation for heavy infantry spears, it should be kept in mind that the Roman cavalry had always continued to employ spears. During the Crisis of the Third Century, however, the Roman heavy cavalry abandoned the traditional model of *lancea* and replaced it with a new one known as the *contus*. This was a type of long wooden lance, originally used by the Persian heavy cavalry. Since the first century AD, this heavy cavalry weapon started to be employed on a massive scale by the Sarmatians. It was later adopted by the Parthian heavy cavalrymen, who employed it against the Romans. The *contus* was about 4m long and had to be wielded by the cavalryman with two hands while directing the horse using his knees, making it a specialist weapon that required a lot of training and good horsemanship to use. Initially, only highly trained cavalrymen such as those fielded by the Sassanids could employ such a weapon effectively. The *contus* was reputedly a weapon of great power, especially if compared to other cavalry spears of its time. The great length of this lance was probably the origin of its name, since the Greek word *kontus* meant 'oar' or 'barge pole'. The Roman cavalry introduced the *contus* on a large scale after facing the Sarmatian heavy cavalry in battle. The light cavalry, instead, continued to be armed for all the period taken into account with normal javelins of the *verutum* model.

Another piece of cavalry equipment copied by the Romans from the Sarmatians was the *draco*. This was a military standard introduced into the Roman cavalry by Sarmatian auxiliaries during the second half of the second century AD. Apparently, it soon proved a great success and was later adopted also by infantry units. In the infantry, it substituted the traditional *vexillum* as the standard of the cohort, meaning the *Vexillifer* of each cohort was replaced by a *Draconarius*. Nevertheless, the eagle

Manica of segmented armour. (*Photograph and copyright by Varianus/Numerus Brittonum*)

Spatha (sword) with its *balteus* (belt) from the mid-third century. (*Photograph and copyright by Conrad Lunar*)

remained the primary symbol of the legions. The *draco* was a traditional element of the Dacian military equipment, but it had been later adopted by the Sarmatians, who subsequently passed its use to the Romans. As suggested by its name, it had the form of a dragon, with open wolf-like jaws containing several metal tongues. The hollow head of the dragon was metal and was mounted on a pole, and it had a long fabric tube affixed at the rear. When in use, the *draco* was held up into the wind, where it filled with air and made a shrill sound as the wind passed through its metal tongues. As a consequence, in addition to being a military standard, it was also a psychological weapon that resembled a terrifying dragon.

During the Early Empire, the Roman Army had always included a large number of light infantrymen equipped as slingers. These were initially mercenaries and later auxiliaries from the Balearic Islands, which had an ancient military tradition that was strongly linked to the use of the sling as a skirmishing weapon. Slings eventually started to be used by soldiers coming from every corner of the Empire, and the original model of sling traditionally employed in the Balearic Islands was joined by a new type. The slingers of Constantine's times were mostly equipped with the staff sling, which this consisted of a staff (a length of wood) with a short sling at one end. One cord of the sling was firmly attached to the staff, while the other end had a loop that could slide off and release the projectile. Staff slings were extremely powerful, with the staff often being made as long as 2m, creating a formidable lever. Ancient art shows slingers holding staff slings by one end with the pocket behind them, using both hands to throw the staves forward over their heads. The staff sling could throw heavy projectiles at much greater distance and at a higher arc than a traditional Balearic hand sling, and in the hands of an experienced skirmisher could be just as accurate.

Archers were always a secondary element of the Roman Army during the Early Empire. The bow had never been an important weapon in the Roman military tradition, and all the units of archers employed by the Empire comprised provincial auxiliaries. With the cavalry reforms of the third century, however, the Romans started to equip part of their cavalry as mounted archers on the model of the Sarmatian and Sassanids horsemen. As a result, the Roman light cavalry adopted the composite bow of the Eurasian steppes as one of its main weapons. At the same time, large numbers of light infantry units started to be defined as *sagittarii* and were equipped with bows of the traditional wooden model employed in continental Europe. The composite bow was made from horn, wood and sinew laminated together. The horn was on the belly, facing the archer, while sinew was on the outer side of the wooden core. The wooden core gave the bow its shape and dimensional stability. When the composite bow was drawn, the sinew (stretched on the outside) and the horn (compressed on the inside) stored more energy than the wood for the same length of bow. The construction of a composite bow was a very complex process, requiring more varieties of material than a wooden bow and much more time. It was often made of multiple pieces, joined with animal glue in V-splices. Pieced construction allowed the use of woods with different mechanical properties for the bending and non-bending sections: indeed, the wood of the bending part of the limb had to endure intense shearing stress. A thin layer of horn was glued onto what would be the belly of the bow; this could store more energy than wood in compression. Goat and sheep horn was commonly used for this purpose. The sinew, soaked in animal glue, was then laid in layers on the back of the bow, the strands of sinew being oriented along the length of the bow. The sinew was normally obtained from the lower legs and back of wild deer or domestic ungulates. Sinew would extend further than wood, again allowing more energy storage. Hide glue was used to attach layers of sinew to the back of the bow and to fix the horn belly to the wooden core. The animal glue could lose strength in humid conditions and be quickly ruined by submersion or rain, so composite bows were always stored in protective leather cases. The wooden bow of the infantry, the *arcubus ligneis*, was constructed according to the longbow tradition of continental Europe from a single piece of wood. Crafting a composite bow could take a week's work, plus a drying time of some months, whereas a wooden bow could be made in a single day and dried in just a week. Peoples living in humid or rainy regions have always favoured wooden bows, while those in dry or arid areas have preferred composite ones. The main advantage of composite bows over longbows was their combination of smaller size with higher power. In addition, almost all composite bows were recurve, with their shape curving away from the archer, a design which gave higher draw-weight in the early stages of the archer's draw, thus storing more total energy. The method used by Roman mounted archers to carry arrows for their composite bow showed a clear nomadic influence, being held pointing upwards in a barrel quiver hanging from the waistbelt. Infantry quivers, instead, were round-bottomed cylinders hanging from a shoulder-strap (which held the arrows point downwards). In either case, each archer normally carried between thirty and forty arrows at a time.

During the Early Empire, Roman battle tactics were highly standardized but incredibly effective. The first weapon to be used in a pitched battle, as we have seen, was the *pilum*. Later, once the enemy ranks had been shattered by the initial rain of javelins, the legionaries drew their short swords and charged their opponents. According to the Roman tactical doctrine, emphasis was put on using the *scutum* to provide maximum body coverage, while the *gladius* was used to attack with devastating thrusts and short cuts. Thanks to such tactics, the Romans were able to defeat any enemy infantry for several centuries, while limiting the number of casualties suffered

Sword of the Illerup Adal model. (*Photograph and copyright by Varianus/Numerus Brittonum*)

by imperial troops. Using their swords to thrust in the few spaces created between the shields of their close formations, the legionaries were rarely exposed to the offensive weapons of the enemies, who had very few chances to manoeuvre. Intensive and continuous training made the Roman legionaries very efficient: the stabbing wounds produced by their short swords were almost always mortal, especially if striking the abdominal area (intended as the main target for thrusts since the early phase of a legionary's intensive training). Each Roman infantryman was trained to adapt to any possible combat situation. Each of his weapons could be used in different ways and he had to be ready to exploit at full any mistake of the enemy or any favourable momentum. For example, Roman legionaries were trained to slash kneecaps beneath their shield wall or to cut the throat of the enemies while charging in the *testudo* formation. In the latter, the legionaries aligned their rectangular shields to form a packed rectangle covered at the front and on the top. The first row of men held their shields from about the height of their shins to their eyes, in order to present a shield-wall to all sides. The men in the rear ranks placed their shields over their heads to protect the formation from above, balancing the shields on their helmets by overlapping them. The legionaries on the sides and rear of the *testudo* held their shields like those of the first row. This formation of the Roman legions offered excellent protection against the missile weapons of any possible opponent.

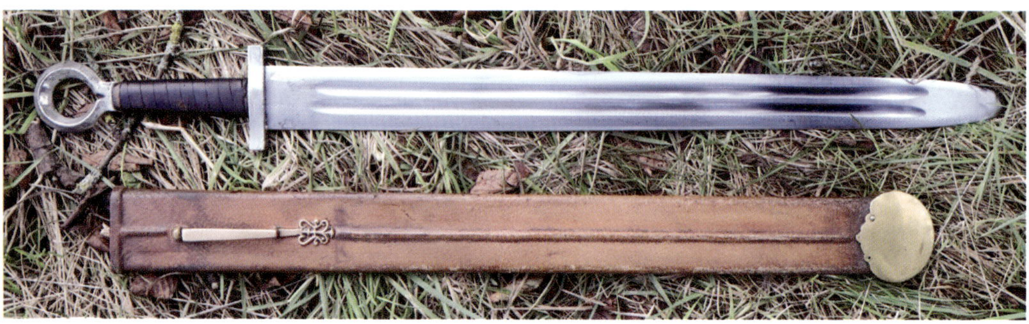

Sword of the Ring Pommel model. (*Photograph and copyright by Varianus/Numerus Brittonum*)

During the glory days of the Early Empire, the Romans usually employed a general strategy that is commonly known as forward defence, with the legions stationed on the frontiers of the Empire prepared to neutralize imminent enemy incursions or invasions before they could reach the territories of Rome. This strategy had proved to be very efficient on several occasions, mainly thanks to the creation of strategic salients beyond the *limes*. This system of defence, however, had been created to handle the low-intensity raids of the Germanic tribes or the invasions of regular Parthian and Sassanid armies. It was not suited to stop mass migrations of entire peoples, like those that occurred on the Rhine and Danube frontiers starting from the reign of Marcus Aurelius. At the end of the second century AD, the Roman military machine was completely surprised by this new kind of threat. The generals of Marcus Aurelius had to struggle for years against the Germanic tribes in order to repulse their invasions, suffering very heavy losses that shattered the basic structure of the legions. After these dramatic events, it soon became clear to the imperial military leaders that a diffused defence of the borders was no longer the right response to foreign threats. As a result, Gallienus and the following rulers started to create large reserves of highly mobile troops that were stationed around the major cities of the Empire. The forward defence strategy was now too vulnerable, as the migrating Germanic tribes could easily concentrate many thousands of warriors in a particular point of the *limes*, which was usually defended by a single legion with its auxiliary units. The lack of reserves to the rear of the border meant that a foreign invading force could penetrate very deeply into the Empire before Roman reinforcements from other border garrisons could arrive to intercept it.

The new defensive strategy created by Gallienus and developed by his successors is commonly known as defence in depth. The Romans accepted that their frontier provinces would become the main combat zone in the operations conducted against the barbarians, rather than the enemy lands located across the border. Whereas the previous forward defence had been based on the principle that the best defence was preventive offence, the new strategy was much more passive and definitively abandoned any possibility of territorial expansion for Rome. This change of general strategy was

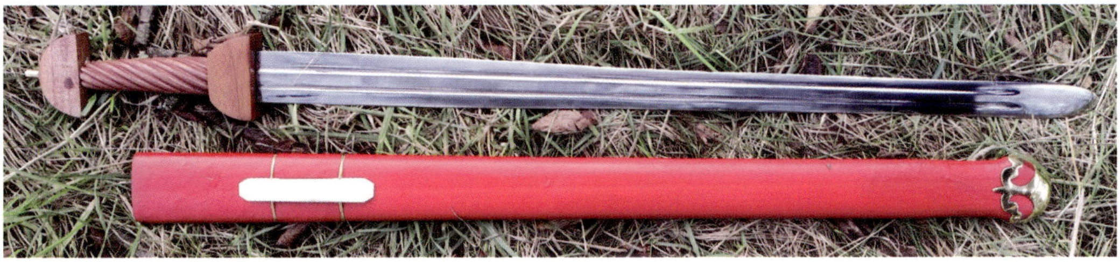

Sword of the Vimose model. (*Photograph and copyright by Varianus/Numerus Brittonum*)

Sword of the Thorsberg model. (*Photograph and copyright by Varianus/Numerus Brittonum*)

completed during the crucial years that saw the great military reforms of Diocletian and Constantine. The creation of *comitatenses* and *limitanei* was fundamental in this sense: the immense borders of the Empire continued to be protected by garrisons, but these were now formed of static troops having purely defensive functions. The *limitanei* did not have the capability to launch preventive offensives against enemies, their main role being to deal with small raids and incursions, which became very common during the continuous low-intensity conflicts happening on the *limes*. In the event of large invasions by migrating peoples, the *limitanei* had to avoid direct confrontation with the enemy. Instead, they just had to keep their defensive positions and structures located on the border, in order to slow down the advance of the invaders. The time gained by the *limitanei* was of fundamental importance for the *comitatenses*, who were usually assembled quite rapidly by the imperial authorities and sent against those who had penetrated the border. As a result of this standard practice, the decisive battle was to be fought inside the Roman province that had been attacked (sometimes in a location that was near to the border, but on other occasions near a major urban centre where the *comitatenses* were garrisoned). The auxiliaries of the Early Empire, according to the previous principle of forward defence, had great offensive capabilities compared with the *limitanei*, being much better equipped and more numerous than the later static troops. As we have seen, among the *comitatenses* forces there was also the *comitatus praesentalis* (the personal army of the emperor). This could be used like all the other field armies to counter foreign invasions, but its main function was to defeat any attempt of rebellion or usurpation directed against the emperor. After the end of the third century, secessions and civil wars remained quite common in the Roman world. The new system of defence was supplemented by a series of alliance treaties that were arranged with the various tribes or client states located on the Roman borders. These treaties of mutual support were based on a very simple principle: the Romans promised to defend their local ally from attacks mounted by its neighbours, while in return, the local ally promised to refrain from raiding the imperial territories.

Regarding the evolution of tactics, the most important change of the third century was the ascendancy of cavalry. For many centuries, since the days of the Greek hoplites who fought against the Persians in the fifth century BC, the armies of the Mediterranean world had seen a clear prominence of the infantry over the cavalry. The Romans had conquered immense territories through the power of their legions, exporting their political control due to the superiority of the legionary heavy infantryman. But the epochal events happening during the Middle Empire changed all this, leading to the progressive decay of the long tradition of infantry warfare. Contact with new military civilizations, like those of the Sarmatians and Sassanids, led the Romans to adopt new tactical principles. These were mostly employed as a response to the new military threats faced since the third century, being strongly linked to the adoption of the new defence in depth strategy. As we have seen, under Gallienus, the number of cavalry increased significantly. Until that time, mounted troops had only been a secondary element of the Roman military apparatus, comprising just small contingents of legionary cavalry by units of auxiliaries. The cavalry troops taking over tactical prominence from the infantry did not only have numerical consequences: the general status of the mounted forces was also deeply affected by the tactical changes. It should be noted that Roman generals of the Middle Empire usually preferred to avoid major pitched battles with large enemy armies if possible, in contrast to what had happened during previous centuries. This was mainly due to the recruiting problems described previously and the economic difficulties of the Empire, which meant that replacing the losses suffered by the elite *comitatenses* armies was difficult to do, especially if these casualties affected cavalry contingents. Producing equipment and breeding horses for the cavalry was very expensive, so losing thousands of *comitatenses* in a single battle was not sustainable because it meant Rome being militarily weak for several years. So the main reasons behind avoiding pitched battles were purely practical ones, and this should not be interpreted as a sign of military inferiority – Roman armies were still able to win most of the field battles fought against barbarian forces during the third and fourth centuries. The Roman military superiority may not have been as marked as in the days of the Early Empire, but if compared with the Germanic tribes or the Sassanids, the elite *comitatenses* retained a more advanced level of training and organization. In order to minimize casualties, the Roman generals had several tactical options at their disposal: night attacks, ambushes, surprise raids, harassment and strategic manoeuvring.

Detail of a sword's suspension system. (*Photograph by Rachele Lori, copyright by Protectores Domini Nostri*)

Pugiones (daggers) used by legionaries during the Crisis of the Third Century. (*Photograph and copyright by Varianus/Numerus Brittonum*)

Tactics employed during pitched battles, however, were not so different from those of the great field clashes during the Early Empire, the main tactical innovations being related to the use of new categories of troops such as mounted archers. The heavy infantry of the legions was drawn up in a main line, which was usually straight and several ranks deep, while light infantry (*auxilia palatina*) and light cavalry (mostly mounted archers) were stationed in skirmishing formation in front of the main infantry line. Heavy cavalry formed the wings, while archers and slingers were deployed to the rear of the heavy infantry line. The disposition of the army was

Gilded head of a *draco* standard. (*Photograph and copyright by Conrad Lunar*)

completed by a large reserve, which had a variable strength and included infantry or cavalry units. This was located to the rear of the main line, in order to deal with breaches opened by the enemy in the heavy infantry formations or to exploit any tactical opportunity that arose. Each army deployed on the field had its camp in the rear, which was guarded by a small garrison and could have different levels of fortification. The camp contained the baggage of the army and thus could act

Nice example of *scorpio*, a kind of light artillery weapon employed by the Roman legions. (*Photograph by Simone Ochsner (https://www.simoneochsner.com), copyright by Legio XI CPF*)

as a refuge in case of rout. Generally speaking, the defences of the camps were of lesser quality and complexity than those built by the armies of Augustus or Trajan. Nevertheless, a ditch was generally dug around the perimeter of the camp and the resulting spoil was used to erect a rampart. The rampart was topped by a palisade of sharpened wooden stakes that were arranged in order to form a defensive screen. In battle, the screens of light infantry and light cavalry would throw their missiles on the enemy before the decisive engagement of the heavy infantry lines; after performing this purely skirmishing function, they would retreat to the rear, where, together with the foot archers and slingers already deployed there, they would continue firing a rain of missiles on the enemy infantry. The task of the heavy cavalry contingents on the wings was to scatter the enemy cavalry facing them and then, if possible, encircle the main body of the enemy infantry. These tactical moves were not so different from those used during the Early Empire, the main differences involving the heavy infantry. Differently from before, the Roman legionaries now acted quite passively during pitched battles, as instead of moving against their enemies as they had done previously, they preferred to await the attack of the enemy infantry by keeping their defensive positions and forming a wall of shields. Thanks to the new combination of oval shield and longer offensive weapons (the *lancea* and *spatha*), the Roman heavy infantrymen tried to use their superior coordination and organization to repulse enemy attacks with crippling losses. When this happened, the battle was won; but if the enemy was able to break the line of Roman heavy infantry, the battle was lost. Missile weapons were much more important than before: during the Early Empire, the *pila* of the legionaries had the same tactical role that was now performed by specific skirmishing lines of light infantry and cavalry. Additionally, the foot archers deployed behind the heavy infantry now played an important role during all the phases of the battle. This general preference for fighting from distance can again be explained by the main strategic need: keeping human and material losses of the *comitatenses* to a minimum. Over time, light troops became increasingly important, with mounted archers, in particular, starting to be much more numerous than cavalrymen equipped with heavy armour.

Bibliography

Primary Sources

Ammianus Marcellinus, *Res Gestae.*
Appian, *Roman History.*
Appianus, *Gallic History.*
Appianus, *Illyrian Wars.*
Appianus, *Wars in Spain.*
Dio, Cassius, *Roman History.*
Diodorus Siculus, *History.*
Diodorus Siculus, *Library of History.*
Dionysios of Halikarnassos, *Roman Antiquities.*
Eusebius of Caesarea, *Chronicon.*
Eusebius of Caesarea, *Historia Ecclesiastica.*
Livy, *History of Rome from its foundation.*
Plutarch, *Lives.*
Polybius, *The Histories.*
Strabo, *Geography.*
Tacitus, *Agricola.*
Tacitus, *The Annals.*
Tacitus, *The Histories.*

Secondary Sources

Anderson, E.B., *Cataphracts: Knights of the Ancient Eastern Empires* (Pen & Sword, 2016).
Baker, P., *Armies and Enemies of Imperial Rome* (Wargames Research Group, 1981).
Bishop, M.C. and Coulston, J.C., *Roman Military Equipment From the Punic Wars to the Fall of Rome (Oxbow Books, 2006).*
Brzezinski, R. and Mielczarek, M., *The Sarmatians 600 BC–AD 450* (Osprey Publishing, 2002).
*Connolly, P., Greece and Rome at War (*Macdonald Phoebus Ltd, 1981).
Cowan, R. and McBride, A., *Imperial Roman Legionary AD 161–284* (Osprey Publishing, 2003).
Cowan, R. and O'Brogain, S., *Roman Guardsman 62 BC–AD 324* (Osprey Publishing, 2014).
Cowan, R. and O'Brogain, S., *Roman Legionary AD 284–337* (Osprey Publishing, 2014).
Elliot, P., *Legions in Crisis: Transformation of the Roman Soldier AD 192–284* (Fonthill Media, 2014).
Elton, H., *Warfare in Roman Europe AD 350–425* (Clarendon Press, 1998).
Esposito, G., *The Late Roman Army* (Winged Hussar Publishing, 2016).
Farrokh, K., *The Armies of Ancient Persia: the Sassanians* (Pen & Sword, 2014).
Goldsworthy, A., *The Fall of the West: The Slow Death of the Roman Superpower (*Weidenfeld & Nicolson, 2009).

Gorelik, K., *Warriors of Eurasia* (Montvert Publishing, 1995).
Kiley, K.F., *An illustrated Encyclopedia of the Uniforms of the Roman World* (Lorenz Books, 2012).
Macdowall, S. and Embleton, G., *Late Roman infantryman 236–565 AD* (Osprey Publishing, 1994).
Macdowall, S. and Hook, C., *Late Roman Cavalryman AD 236–565* (Osprey Publishing, 1995).
Macdowall, S. and McBride, A., *Germanic Warrior AD 236–568* (Osprey Publishing, 1996).
Newark, T., *Ancient Armies* (Concord Publications, 2000).
Newark, T., *Ancient Celts* (Concord Publications, 1997).
Newark, T., *Barbarians* (Concord Publications, 1998).
Newark, T., *Warlord Armies* (Concord Publications, 2004).
Nicasie, M., *Twilight of Empire: The Roman Army from the Reign of Diocletian until the Battle of Adrianople* (Brill Academic Publishers, 1998).
Nicolle, D. and McBride, A., *Romano-Byzantine Armies 4th–9th centuries* (Osprey Publishing, 1992).
Nicolle, D. and McBride, A., *Rome's Enemies 5: the Desert Frontier* (Osprey Publishing, 1991).
Quesada Sanz, F., *Armas de Grecia y Roma* (La Esfera, 2014).
Rankow, B. and Hook, R., *The Praetorian Guard* (Osprey Publishing, 1994).
Simkins, M. and Embleton, R., *The Roman Army from Hadrian to Constantine* (Osprey Publishing, 1979).
Sumner, G., *Roman Military Clothing (1): 100 BC–AD 200* (Osprey Publishing, 2002).
Sumner, G., *Roman Military Clothing (2): AD 200–400* (Osprey Publishing, 2003).
Syvanne, I., *Military History of Late Rome 284–361* (Pen & Sword, 2014).
Travis, J. and Travis, H., *Roman Body Armour* (Amberley, 2012).
Travis, J. and Travis, H., *Roman Helmets* (Amberley, 2015).
Travis, J. and Travis, H., *Roman Shields* (Amberley, 2014).
Treadgold, W., *Byzantium and Its Army, 284–1081* (Stanford University Press, 1995).
Warry, J., *Warfare in the Classical World* (Salamander Books, 1997).
Wilcox, P., *Rome's Enemies 1: Germanics and Dacians* (Osprey Publishing, 1982).
Wilcox, P., *Rome's Enemies 2: Gallic and British Celts* (Osprey Publishing, 1985).

The Re-enactors who Contributed to this Book

Ala I Batavorum

Some clouds of dust on the horizon, slowly heading your direction ... It turns out to be a squad of Roman cavalrymen, each fully armed, with shiny armour and a highly decorated horse. The approaching horsemen wear silver masks, blinking in the sun, their faces covered, only showing their eyes: spectacular, yet terrifying. Stichting Ala I Batavorum recreates these cavalrymen from ages gone by. In our regular shows, our fully equipped cavalrymen display their skills at arms on horseback. Equipment and tactics will be shown and explained, both in single file as well as operating in the group, allowing all equipment and the tactical advantage of the cavalry within the Roman Army to be explained to the public. The group provides great entertainment: a spectacular and educational show during your event that will have your guests talking about it for some time. Stichting Ala I Batavorum can bring together up to six cavalrymen. All our equipment is based upon the archaeological records and well researched. Our impressions focus mainly on the Batavian *auxilia* cavalry, but also include Augustean and Late Roman. Most impressive is our late Roman *clibinarius* kit, incorporating a recreation of the Dura Europos (Syrian) horse scale armour.

Contacts:
E-mail: info@ala-batavorum.nl
Website: www.ala-batavorum.nl
Facebook: https://m.facebook.com/ALA-BATAVORUM-133726753365976/

Conrad Lunar

Conrad Lunar is a German re-enactor living in Neuwied in northern Rhineland-Palatinate. Nieberbieber is part of Neuwied, where the largest castle of the upper-Germanic *limes* is located. Some unique findings were made there, especially the Draco and the Niederbieber helmet type. Conrad Lunar shows a typical Roman soldier of the late second and third century. His project in Neuwied is known as www.Drachenkopfweg.de – a 42km trail around Neuwied, where you can explore the Roman history from Caesar (in this area he built the very first bridges across the

Rhine), the UNESCO World Heritage borders of the Roman Empire and a Late Roman harbour – more than 400 years of Roman history in a small area.

Contacts:
Facebook: https://www.facebook.com/profile.php?id=100014947538481
Website: www.Drachenkopfweg.de

Legio VI Victrix Cohors II Cimbria

Legio VI Victrix Cohors II Cimbria is a Scandinavian Roman re-enactment group started in Denmark, but today also with a chapter in Sweden and members from Germany. Our group consists of both military and civilian re-enactors, reproducing life in the Roman legions in the late first century AD. We also convey the life of the Roman Empire and its relations with the regions north of Germania Inferior, the Cimbrian peninsula and Scandia. Every summer, Cohors II Cimbria participates in events both in Scandinavia and the rest of Europe. In the off-season we research, lecture and are constantly developing our knowledge and equipment. Roman re-enactment is a never-ending process, as new archaeological finds and fresh theories keep appearing and constantly change the view of the Roman armies. Scandinavia was never a part of the Roman Empire, but due to our many bog findings, most of which are from the Roman period 0–AD 350, we have in the region a lot of Roman artefacts. Among these is the largest collection of Roman imperial weapons in the world outside Italy. The tribes that lived here at that time – mainly the Cimbrians, Teutons, Charudes and Sviones – were, from the time of Augustus onward, apparently on friendly terms with the Romans. The *Res Gestae* of Augustus, a primary source, mentions an expedition to these most northern regions in AD 5, led by Tiberius. That is one of many reasons why we find it right and proper to perform a Roman re-enactment in Scandinavia. To us, there is a forgotten story to be told.

Contacts:
E-mail: cohorsllcimbriadk@gmail.com
Website: https://cohorsllcimbria.com/
Facebook: https://www.facebook.com/Cohorsllcimbria/

Legio XI CPF

Our association, the Vex Leg XI CPF (Switzerland), exists since November 2006. The abbreviation stands for 'Vexillatio Legionis Undecimae Claudiae Piae Fidelis' (i.e. 'Detachment of the XI Claudian Pious and Loyal Legion'). The title 'Loyal' is a tribute awarded under Emperor Claudius for loyal duty. Our club considers itself part of a worldwide movement called 'living history'. We deal with Roman military history especially in the era of the classical imperial period, during the second half of the first century AD. We try to depict everyday life in a military environment as realistically as possible. The only Legionary Camp in Switzerland was located in Vindonissa (present-day Windisch in the Canton of Aargau). During its existence, three legions served there, the last of which was the Legio XI. As Swiss citizens, taking one of these three as the subject of our reconstructions was the obvious choice. We chose the XI Legion because it rebuilt the camp in stone and was stationed there for thirty years, longer than any other legion serving in Vindonissa. After the turmoil of the so-called 'Year of the Four Emperors' (AD 69/70), Emperor Vespasian decided to move the LEGIO XI CPF to Vindonissa in order to replace the LEGIO XXI that had cruelly suppressed an uprising of the Helvetii. The service of the Legio XI CPF in the legionary camp of Vindonissa is documented for the period from AD 70 to AD 101. During this time, the legionary camp was expanded into a permanent one and the military and logistical infrastructure along the transalpine route in southern Germany was expanded. During this time, the LEGIO XI CPF was deployed by Emperor Domitian to fight in the province of Germania Superior in the Chatti Wars. Under Emperor Trajan, in AD 101, the LEGIO XI was withdrawn from Vindonissa and stationed in Brigetio (Komorn) and in Aquincum (Budapest) in Pannonia in order to take part in the Dacian Wars. As an association with the aim of conveying history, we work closely with museums throughout Switzerland and abroad at various events. Of course, this also includes visiting schools as part of regular lessons and even supporting high school exams. Roman legionnaires work together with the teachers to make the lessons exciting and practical. This can include reconstructing military and civil aspects of Roman life. All the members strive to implement the function they embody to the best of their ability. This is the main goal defined by our club. That is why we continually engage with relevant specialist literature, exchange information about current research or new results and adapt our equipment and items of use if necessary. However, this playful representation of history is by no means inferior to established findings. It is simply another way of presenting facts and figures. We try to provide a clear picture of the current state of knowledge. We have also been able to participate in projects under the supervision of archaeologists and experts regarding the replacement of equipment and its functionality. We rely on the persuasive power of what is real, tangible and comprehensible. We are therefore not afraid of getting in touch with today's people and anyone interested is always welcome to exchange knowledge and stimulating discussions. Our club life is characterized by sociability and friendship. Social events, such as shared meals or club trips, are just as important to us as the presentation of history.

Contacts:
E-mail: info@legioxi.ch
Website: https://legioxi.ch/
Facebook: https://www.facebook.com/profile.php?id=100081285702381

Numerus Brittonum

The Numerus Brittonum (NBL) bring to life the military and civilian context of a border society on the Upper Germanic-Rhaetian *limes* of the third century AD. We were founded in 2005 as a division of the Historical Association Welzheimer Wald e.V. and are based in Welzheim-Eastern-Fort. We offer an internal and external programme that balances public events and exclusive activities, with the goal of making history fun and accessible. We aim to provide a new dimension to the archaeological sites at the cultural heritage *limes*, and to contribute to their significance and protection. We focus on the revival of a Roman auxiliary unit of the third century AD, under the reign of the Severan emperors, the conveyance of Roman cultural history, preservation of Roman heritage and its visible evidence, such as the Eastern-Fort and the Archaeological Park Welzheim. During our time at the *limes*, we are stationed at the Eastern-Fort in Welzheim, which housed around 200 soldiers. There is a smaller Western-Fort for approximately twenty soldiers nearby. We also have a *vicus*, a village responsible for the supply and accommodation of civilian personnel. Our equipment and clothes are based on the time period of approximately AD 160–260. The NVMERVS BRITTONVM is dedicated to bringing Roman history to life and providing an immersive experience for visitors and members alike. Our commitment to authenticity, pedagogical conveyance and bridging the gap between modernity and the past sets us apart in the world of re-enactment. We are most renowned for our fun programmes for children and provide the best display of third-century Romans worldwide. We are always open to new members; more information can be found on our website.

Contacts:
E-mail: Nbl-vorstand@list.numerus-brittonum.de
Website: www.numerus-brittonum.de and www.lebenamlimes.de

Paul Elliott

Paul Elliott has been involved in Roman re-enactment since 2004, previously gaining an honours degree in Ancient History & Archaeology. His books have combined experimental archaeology with historical research, and include *The Last Legionary*, *Legions in Crisis*, *Warfare in New Kingdom Egypt*, *Everyday Life of a Soldier on Hadrian's*

Wall and *Food and Farming in Prehistoric Britain*. He has an extensive collection of reproduction Roman, Bronze Age and Iron Age military objects, and frequently works in a local school and at a heritage museum in the north of England. He and his fellow re-enactors are frequent visitors to Rome and to Roman sites across Britain.

Contacts:
E-mail: paulelliottbooks@yahoo.com
Website: https://www.paulelliottbooks.com/history-books.html

Protectores Domini Nostri

The group Protectores Domini Nostri aims to bring to life the personal guard, *Concilium Principis* and Imperial *Comitatus* of a Roman emperor in the period from the end of the third century AD to the beginning of the fourth century, with particular reference to the reign of Emperor Maxentius (AD 306–312). Through the study of literary sources, iconography and the interpretation of archaeological finds, the group attempts to reconstruct the military and ceremonial aspect of the restricted imperial entourage of this specific time frame. The *Protectores*, the name of the chosen soldiers who were selected from among the best elements of the army ranks to serve under the Augusti, were presumably all former officers, and through their service at court they were projected to embark on a career that would lead them to the highest posts. The name chosen for the group originates from a funerary inscription found in Rome, where a certain Valerius refers to himself in the inscription precisely as '*PROTECTOR DOMINI NOSTRI MAXENTI AVGVSTI*'. In addition to the *Protectores*, the group has also reconstructed a Late Imperial legion, that of the *Moesiaci Seniores*, a unit attested at the time of Maxentius in the north-eastern Italian fortresses of Aquileia and Verona. The president of the group, Marco Cecini, attempted to reconstruct the figures of the Emperors Maxentius and Aurelian (AD 270–275), with a thorough investigation of the available iconographic and literary sources, especially numismatic ones, which are essential for a correct understanding of the imperial clothing of the period of interest. The group has collaborated with a local Italian television station, Canale 10, on the production of popular historical content, and is working on the completion of a self-produced documentary on the life of Emperor Maxentius.

Contacts:
Email: protectoresaugusti@gmail.com
Facebook: https://www.facebook.com/ProtectoresAugusti/

Varianus

Welcome to the captivating world of VARIANVS, where history becomes reality! Immerse yourself in the dedication and passion of two soldiers from the third century AD as you explore this unique form of storytelling in our presence. Experience the meticulous attention to detail and commitment to authenticity that goes far beyond just 'dressing up'. 'How did they do it?' 'What materials did they use?' From small details like how they tied their belts to understanding the behaviour of the Roman military, every element is carefully researched and crafted to transport you back in time. Our goal is to display only evidenced equipment, with as little speculation as possible to complete the display. VARIANVS are not only passionate re-enactors available for events, but also actors for films, documentaries and other projects. We are willing to use client-provided equipment to ensure seamless integration into various productions. By summarizing the world of re-enactment, www.varianvs.com serves as a bridge between audiences and a topic they may not have explored otherwise. We lower the threshold for accessing historical ideas, condensing hours of research into easily digestible minutes. Join us at multiple events throughout the year in and around Germany. You can find an event calendar on our website. Be sure to catch us at every event with the best Roman third-century group worldwide – NVMERVS BRITTONVM (www.numerus-brittonum.de). VARIANVS invites you to step into the realm of re-enactment and witness history unfold before your eyes. Our dedication to authenticity, availability as actors and commitment to bridging the gap between modernity and the past, as well as our unique stage combat show, make us a unique presence in the world of re-enactment.

Contacts:
E-mail: official@varianvs.com
Website: https://varianvsofficial.wixsite.com/website

Index

Antonine Plague, 2, 102
Antonine Wall, 57, 59

Bagaudes, 32
Burgarii, 101

Castellani, 101
Constans, 87

Didius Julianus, 53
Dilectus, 106

Fabricae, 152

Isauria, 93

Laeti, 102–103
Licinius, 73, 75, 83, 87, 89

Macrinus, 61–2
Maximinus Daza, 71, 75, 79
Maximinus Thrax, 62
Mediolanum, 32, 40, 75, 83

Palmyra, 6, 23, 39, 66–7, 88
Pertinax, 53
Pescennius Niger, 53
Postumus, 6, 66, 67
Probus, 71

Riparienses, 101

Scorpio, 21, 120
Septimius Severus, 2, 6, 45, 53–4, 57, 61
Sirmium, 32, 40

Testudo, 178
Tirones, 106
Tunica manicata, 148, 154

Valerian, 6, 66
Vitis, 149

Zenobia, 67